Environmental Methods for Transport Noise Reduction

Environmental Methods for Transport Noise Reduction

Edited by Mats E. Nilsson
        Jörgen Bengtsson
        Ronny Klæboe

CRC Press
Taylor & Francis Group
Boca Raton London New York

CRC Press is an imprint of the
Taylor & Francis Group, an **informa** business

A SPON PRESS BOOK

CRC Press
Taylor & Francis Group
6000 Broken Sound Parkway NW, Suite 300
Boca Raton, FL 33487-2742

First issued in paperback 2019

© 2015 by Taylor & Francis Group, LLC
CRC Press is an imprint of Taylor & Francis Group, an Informa business

No claim to original U.S. Government works

ISBN-13: 978-0-415-67523-9 (hbk)
ISBN-13: 978-0-367-86625-9 (pbk)

Library of Congress Cataloging-in-Publication Data

Environmental methods for transport noise reduction / editor[s] Mats Nilsson, Jorgen Bengtsson, and Ronny Klaeboe.
   pages cm
   Includes bibliographical references and index.
   ISBN 978-0-415-67523-9 (alk. paper)
   1. Transportation noise. 2. Railroads--Noise. 3. Noise barriers. 4. Plants as noise barriers. 5. Trees. I. Nilsson, Mats, 1964- editor. II. Bengtsson, Jorgen, editor. III. Klaeboe, Ronny, editor.

TD893.6.T7E58 2015
620.2'3--dc23   2014016895

Visit the Taylor & Francis Web site at
http://www.taylorandfrancis.com

and the CRC Press Web site at
http://www.crcpress.com

Contents

7 Porous ground, crops, and buried resonators   153

KEITH ATTENBOROUGH, SHAHRAM TAHERZADEH, IMRAN BASHIR,
JENS FORSSÉN, BART VAN DER AA, AND MANUEL MÄNNEL

8 Vegetation in urban streets, squares, and courtyards   177

JIAN KANG, MAARTEN HORNIKX, TIMOTHY VAN RENTERGHEM,
YULIYA SMYRNOVA, JENS FORSSÉN, CHRIS CHEAL, DICK BOTTELDOOREN,
HONG-SEOK YANG, JIN YONG JEON, HYUNG SUK JANG,
SHAHRAM TAHERZADEH, KEITH ATTENBOROUGH, AND AGNES MANDON

9 Perceptual effects of noise mitigation   195

MATS E. NILSSON, DICK BOTTELDOOREN, JIN YONG JEON,
MARIA RÅDSTEN-EKMAN, BERT DE COENSEL, JOO YOUNG HONG,
JULIEN MAILLARD, AND BRUNO VINCENT

Preface

Exposure to noise from roads and railways is widespread, and the problem is increasing, primarily as a consequence of the continuous urbanization and growth of the transport sector. Traffic noise causes annoyance and sleep disturbance, and it interferes with rest, concentration, speech communication, and learning. There also is increasingly strong support for a causal link between long-term exposure to road traffic noise and cardiovascular disease, including hypertension and myocardial infarction.[1]

The most effective noise-mitigation method is to reduce noise emissions at the source, for example, by means of regulations demanding quieter engines, tires, or road surfaces, or by limiting traffic flow volumes and introducing stricter speed limits. However, such methods are often difficult to implement for economic, city planning, or political reasons. Therefore, at-source noise reduction must be complemented with methods that act on the noise during its path to the receiver. The aim of this book is to encourage the use of new and environmentally friendly methods of this kind.

Environmental Methods for Transport Noise Reduction presents the main findings of the research project HOlistic and Sustainable Abatement of Noise by optimized combinations of Natural and Artificial means (HOSANNA). The project aimed to develop a toolbox for reducing road and rail traffic noise in outdoor environments by the optimal use of vegetation, soil, and other natural and recycled materials, in combination with artificial elements.

The HOSANNA project studied a number of abatement strategies that might achieve cost-effective improvements using new barrier designs; planting of trees, shrubs, or bushes; ground and road surface treatments; and greening of building facades and roofs. Vegetated areas and surfaces are greatly appreciated in both urban and rural environments. The beneficial effects of greening mean that the costs of new greening or of maintaining existing green surfaces are often easy to justify, even without considering the benefit of environmental noise reduction. The thrust of the HOSANNA project was to find better ways of using vegetated surfaces and recycled materials to reduce road and rail traffic noise and improve the perceived sound environment. The noise reduction was assessed in terms of sound level reductions, perceptual effects, and cost–benefit analyses.

Traffic noise situations are often complex and a single noise mitigation measure is seldom sufficient. Some of the options we discuss in this book each lead only to 2 to 3 dB(A) reduction in noise, so an appropriate combination of measures is needed to obtain a larger effect. Other individual noise abatements are expected to reduce noise by 10 dB(A) or more. It should be noted that most of the estimated noise reductions have been calculated using advanced numerical methods, rather than measured in real situations, so a nonnegligible uncertainty is expected in real situations. To minimize this uncertainty, the estimation methods have all been validated and are applied in situations that are as realistic as possible. In addition, the impairment in performance due to meteorological effects has been estimated for selected cases by modelling the effects of mean wind and turbulence.

The methods presented in this book act by exploiting various acoustic phenomena that influence sound during their paths from source to receiver. Chapter 1 (Forssén et al.) reviews the general principles of outdoor noise propagation, and specifically those phenomena that are relevant for the efficiency of the mitigation methods, which are introduced in Chapters 2 to 8.

The conventional noise control solution is to erect noise barriers, and much has been learned over the years about noise barrier design.[2] However, there is still room for new ideas, as is evident in Chapter 2 (Defrance et al.), where solutions like low-height vegetated barriers and vegetated barrier caps are discussed.

Chapter 3 (Horoshenkov et al.) presents detailed analyses of the acoustic performance of plants and soil, and illustrates how the acoustic absorption of soils can be enhanced by selecting the right type of low-growing plants. Chapter 4 (Van Renterghem et al.) presents corresponding results for hedges, trees, and tree belts, and their effect on reflection, diffraction, and scattering of sounds. Chapter 5 (Van Renterghem et al.) provides design tips for planting trees and tree belts along roads. Planting schemes may take advantage of several acoustic phenomena, such as multiple scattering in tree belts and upward refraction by trees planted close to noise barriers.

Sound travelling directly from source to receiver will interact with sound reflected from the ground, a phenomenon called *ground effect*. Chapter 6 (Attenborough et al.) suggests a new set of noise control options that uses the ground effect. Examples are the distribution of small protruding elements or grooves over the ground in such a way that the ground effect cancels sound in a frequency range that will reduce the noise from surface transport. Chapter 7 (Attenborough et al.) follows this up by discussing how different ground types give rise to different ground effects, and how this knowledge can be used to choose grounds for improved noise reduction. This chapter also includes a section on how to improve the noise-reducing potential of porous asphalt by burying resonating chambers and resonators, which act on a specific frequency region of the noise.

Chapter 8 (Kang et al.) shows how vegetation on facades and roofs can improve the acoustic environment in urban streets, squares, and courtyards, in addition to the aesthetic and ecological benefits of increasing the amount of greenery in the city. Although the acoustic effect of single measures, such as vegetation on a single facade, may be small, combined measures may lead to substantial noise reduction.

The main part of this book discusses noise reduction in terms of sound pressure levels. This gives a fair indication of the corresponding improvement of the perceived acoustic environment. However, noise mitigation also changes the frequency composition and variability of the mitigated noise at the listener location, and may influence the audibility of other sounds in the environment as well as changing visual features of the environment. Such perceptual effects of noise mitigation are discussed in Chapter 9 (Nilsson et al.). Chapter 10 (Klæboe and Veisten) takes evaluation a step farther, and presents economic analyses of noise mitigation measures, using as examples several of the measures proposed in the previous chapters. In the analyses, costs and benefits of a noise mitigation project are valuated and the project is considered cost efficient if it cost less than the total value of the benefits. These analyses show that many of the proposed methods have the potential of being cost efficient, in several cases robustly so.

Mats E. Nilsson, Jörgen Bengtsson, Ronny Klæboe (Editors), and
Jens Forssén (HOSANNA project leader)
On behalf of the HOSANNA project

REFERENCES

1. WHO. 2011. *Burden of disease from environmental noise.* Copenhagen World Health Organization Regional Office for Europe.
2. Kotzen, B., and C. English. 2002. *Environmental noise barriers. A guide to their acoustic and visual design,* 2nd ed. Oxford, UK: Spon Press.

Glossary

Absorbent materials Sound absorbents or absorbing materials reduce the reflection of sound as a result of being porous so that air particle motion associated with sound is able to penetrate and its energy is converted into heat by friction with the walls of the pores.

Absorption coefficient Result of measuring the sound-absorbing property of a surface, usually frequency and angle dependent. The measurement is made at normal incidence in an impedance tube or at random incidence in a reverberation chamber.

Absorption of sound The process by which sound energy is converted to heat. This can happen in the atmosphere through air absorption, nonporous boundary friction or interaction with a porous boundary.

Acoustically hard/soft A surface that reflects all of the sound that arrives at it is described as acoustically hard, whereas a surface that absorbs some or all of the sound that arrives at it is called acoustically soft.

Atmospheric turbulence Random irregular motion or fluctuation in temperature of fluid (e.g., air) induced by wind friction with the ground or by uneven surface heating. It scatters sound to an extent that increases with frequency. In the atmosphere, it reduces ground effects and the acoustical performance of barriers.

Auralisation A method of simulating a real (e.g., an outdoor) hearing experience in a laboratory or through a virtual environment.

Benefit–cost ratio The ratio between the cash value of benefits accruing from a (noise reduction) action and the costs of implementing the action.

Berm An earthen barrier or bank of earth that may be used for noise control. Frequently, berms are made from soil removed during associated construction activities and planted to improve appearance.

Damping ratio A dimensionless measure of how rapidly oscillations decay.

Diffraction The physical phenomenon by which sound bends around the edges of an obstacle, e.g., the top of a noise barrier.

Diffraction grating A regularly spaced array of obstacles to a sound wave that causes enhanced reflection or cancellation when the wavelength, spacing, and angle satisfy certain conditions.

Diffuse A sound field at a receiver is considered to be diffuse if it contains components travelling in all directions.

Drag Drag (sometimes called *air resistance*) is a type of friction that results in forces acting opposite to the relative motion of any object moving with respect to a surrounding fluid.

Drag coefficient The drag coefficient is a dimensionless quantity that is used to quantify the drag or resistance experienced by an object moving in a viscous fluid.

EA *See* excess attenuation.

Excess attenuation Attenuation of outdoor sound in excess of that due to wavefront spreading and, possibly, air absorption.

Flow resistivity A measure of the ease with which air can pass in and out of a porous surface. Specifically, it is given by the ratio of applied pressure gradient to resulting volume flow per unit thickness of material.

Geometric spreading The physical phenomenon by which sounds spread from a source after generation. This means that sound levels will reduce from distance alone. Spherical spreading and cylindrical spreading are special cases giving rise to 6 dB and 3 dB reduction per doubling of distance, respectively.

Ground effect The physical phenomenon (interference) through which sound reflected from the ground and travelling to a receiver along the reflection path either reinforces or cancels sound that arrives at the receiver directly.

Impedance The ratio of pressure to normal velocity at a surface.

Impedance tube A rigid tube with a loudspeaker at one end and an acoustically hard termination at the other, along which it is possible to measure the pressure profile or the complex pressure (i.e., both magnitude and phase) at two or more fixed microphone positions or continuously using a probe microphone.

Insertion loss The insertion loss due to a mitigation measure is the difference between the sound levels at a given location without and with a mitigation measure. Usually stated in decibels (dB).

Insolation Amount of sunlight incident on a surface.

Leaf area density Leaf area per unit volume (can be one-sided or two-sided).

Loudness The perceived intensity of sounds (unit: sone). Also the output of a psychoacoustic model of the perceived loudness of sounds.

Loudness level The loudness of a sound, expressed as the level of an equally loud 1-kHz tone (unit: phon). Also, the output of a psychoacoustic model of the perceived loudness of sounds.

Notice event An auditory event that is noticed by a listener in a given environment.

Open porosity Volume fraction of interconnecting pores that open to the surface of a material.

Porosity   Total fraction of a material occupied by pores including "dead end" ones.

Porous asphalt   An asphalt mix of stones and binder in which a gap in the stone size distribution is deliberately created so as to result in air-filled voids.

Pressure resistance   *See* flow resistivity.

Pressure resistance coefficient   *See* flow resistivity.

Reflection   The process by which the sound incident on a surface is directed away from the surface. During specular reflection, the sound is directed away from the surface at the same angle from the surface as that made by the incident sound. Reflection represents a special form of scattering when the scattering object is very large compared with the incident wavelength.

Reflection coefficient   The fraction of incoming sound intensity that is reflected.

Refraction   The process involving change of sound speed by which the direction of sound penetrating a surface or region is changed.

Resonator   A structure that resonates. If an undamped structure is vibrated at the frequency of resonance (resonant frequency), the amplitude of vibration grows arbitrarily large. Typical resonators include damping and can be used to absorb sound near the resonance frequency.

Reverberant room   Sometimes called a *reverberation chamber*, a room specially constructed with acoustically-hard surfaces, non-parallel walls, and aids to diffusion.

Scattering   The process by which an obstacle influences incident sound. It depends on the relative size of the obstacle compared to an incident wavelength. If the obstacle is very small compared with the wavelength, its influence is small, but the combined influence of multiple scattering may be significant if there is a large number of small obstacles per unit volume.

Scattering coefficient   The fraction of incoming sound power that is scattered.

Sonic crystal   A regularly spaced array of (usually acoustically hard) scattering objects giving rise to stop and pass bands in acoustic transmission at frequencies that depend on the centre-to-centre spacing.

Soundscape   The overall acoustic environment, including sounds from all audible sources.

Specular reflection point   The position on a reflecting surface at which the angle of incidence is equal to the angle of reflection.

Substrate   An underlying layer (a substratum). Material on which plants grow or are attached.

Substratum   *See* substrate.

Surface wave   A wave in the close vicinity of the ground surface characterized by cylindrical spreading and exponential decay with the height above the surface.

Thermal dissipation   Conversion of mechanical energy to heat. Inside a pore of a porous material it accompanies heat transfer between compressions and rarefactions of the pore fluid and pore walls during the passage of a sound wave.

Tortuosity   A measure of the deviation of streamline flow from a straight line through a porous material.

Transfer function   The ratio of signals at two positions in a signal processing chain.

Transfer matrix approach   A method of modelling sound propagation through a layered system in which the velocities or pressures at each interface are included in a matrix.

Viscous loss   Conversion of mechanical energy into heat through fluid viscosity.

Contributors

Keith Attenborough
Engineering and Innovation
The Open University
Milton Keynes, United Kingdom

Imran Bashir
College of Engineering
Maths and Physical Sciences
University of Exeter
Exeter, United Kingdom

Haidj Benkreira
School of Engineering, Design
and Technology
University of Bradford
West Yorkshire, United Kingdom

Dick Botteldooren
Department of Information
Technology
Ghent University
Ghent, Belgium

Chris Cheal
School of Architecture
University of Sheffield
Sheffield, United Kingdom

Bert De Coensel
Department of Information
Technology
Ghent University
Ghent, Belgium

Jérôme Defrance
Centre Scientifique et Technique
du Bâtiment (CSTB)
Marne de Vallée, France

Jens Forssén
Applied Acoustics
Chalmers University of
Technology
Gothenburg, Sweden

Toby J. Hill
Engineering and Innovation
The Open University
Milton Keynes, United Kingdom

Joo Young Hong
Department of Architectural
Engineering
Hanyang University
Seoul, South Korea

Maarten Hornikx
Department of the Built
Environment
Eindhoven University of
Technology
Eindhoven, The Netherlands

Kirill Horoshenkov
Department of Mechanical
Engineering
University of Sheffield
Sheffield, United Kingdom

Hyung Suk Jang
Department of Architectural
Engineering
Hanyang University
Seoul, South Korea

Philippe Jean
Centre Scientifique et Technique
du Bâtiment (CSTB)
Marne de Vallée, France

Jin Yong Jeon
Department of Architectural
Engineering
Hanyang University
Seoul, South Korea

Jian Kang
School of Architecture
University of Sheffield
Sheffield, United Kingdom

Amir Khan
Bradford Centre for Sustainable
Environments
University of Bradford
West Yorkshire, United Kingdom

Tor Kihlman
Applied Acoustics
Chalmers University of
Technology
Gothenburg, Sweden

Ronny Klæboe
Institute of Transport Economics
(TOI)
Oslo, Norway

Faouzi Koussa
Centre Scientifique et Technique
du Bâtiment (CSTB)
Marne de Vallée, France

Wolfgang Kropp
Applied Acoustics
Chalmers University of
Technology
Gothenburg, Sweden

Julien Maillard
Centre Scientifique et
Technique du Bâtiment
(CSTB)
Marne de Vallée, France

Agnes Mandon
Canevaflor®
Tarare, France

Manuel Männel
Müller-BBM
Munich, Germany

Mats E. Nilsson
Gösta Ekman Laboratory
Department of Psychology
Stockholm University
Stockholm, Sweden

Maria Rådsten-Ekman
Gösta Ekman Laboratory
Department of Psychology
Stockholm University
Stockholm, Sweden

René Rohr
Canevaflor®
Tarare, France

Yuliya Smyrnova
School of Architecture
University of Sheffield
Sheffield, United Kingdom

Shahram Taherzadeh
Engineering and Innovation
The Open University
West Yorkshire
United Kingdom

Bart Van der Aa
Applied Acoustics
Chalmers University of
Technology
Gothenburg, Sweden

Timothy Van Renterghem
 Department of Information
   Technology
 Ghent University
 Ghent, Belgium

Knut Veisten
 Institute of Transport Economics
   (TOI)
 Oslo, Norway

Bruno Vincent
 Acoucité
 Lyon, France

Hong-Seok Yang
 School of Architecture
 University of Sheffield
 Sheffield, United Kingdom

The HOSANNA project

This book is based on research conducted in the research project **HO**listic and Sustainable **A**batement of **N**oise by optimized combinations of **N**atural and **A**rtificial means (HOSANNA). The project aimed to develop a set of tools for reducing road and rail traffic noise in outdoor environments by the optimal use of vegetation, soil, and other natural and recycled materials in combination with artificial elements.

The project studied a number of green abatement strategies that might achieve cost-effective improvements using new barrier designs; planting of trees, shrubs, or bushes; ground and road surface treatments; and greening of building facades and roofs. The noise reduction was assessed in terms of sound level reductions, perceptual effects, and cost–benefit analyses.

The project was coordinated by Chalmers University of Technology in Gothenburg, Sweden (coordinator Associate Professor Jens Forssén), and involved 13 partners from 7 countries. The research received funding from the European Union Seventh Framework Programme (FP7/2007–2013) under grant agreement no. 234306, collaborative project HOSANNA.

MÜLLER-BBM

HOSANNA partners: Chalmers University of Technology (Sweden), CSTB (France), Canevaflor (France), IBBT Ghent University (Belgium), Müller-BBM (Germany), Open University (United Kingdom), City of Stockholm (Sweden), Institute of Transport Economics (TOI) (Norway), University of Sheffield (United Kingdom), University of Bradford (United Kingdom), Stockholm University (Sweden), Acoucité (France), and Hanyang University (South Korea).

Chapter 1

Introduction to traffic noise abatement

Jens Forssén, Wolfgang Kropp, and Tor Kihlman

Abstract. Given that noise from surface transport (i.e., road and rail) adversely affects health and well-being, this chapter briefly describes the main sources of such noise, the general principles of outdoor sound propagation, the basics of noise reduction, and the link between noise indicators and perception. To bridge the gap between existing noise exposure levels and current guideline values, which are well supported by health research, we suggest using a variety of measures, considering noise abatement both at source and during propagation. A mix of classical and novel noise control engineering approaches, including soundscaping and incorporating long-term, large-scale perspectives on urban and rural planning, is needed to enable cost-effective abatement measures. The source strengths (i.e., emissions) and their reduction are characterized in this chapter, focussing on the dominant road traffic noise. We describe the road traffic source model used for calculations in the HOSANNA research project. In this model, the spectral properties of the source strengths are exemplified, as is the dependence on driving speed. In reviewing propagation effects, the topic is subdivided into distance, medium, and boundary—the difference in sound level decay over distance between line and point sources is explained; the effects of refraction (i.e., sound path curving under conditions of wind or height-varying temperature) are characterized; the effects of atmospheric absorption and turbulence scattering are described; the ground effect is explained and demonstrated for asphalt-paved and grassed surfaces; and the performance of noise barriers is described using the example of diffraction from a thin rigid barrier.

1.1 BACKGROUND

Sufficient evidence from large-scale epidemiological studies indicates that environmental noise adversely affects human health and well-being.[1] The social costs of traffic noise are significant. The World Health Organization (WHO) has estimated the burden of traffic-related disease in the European Union (EU) as a loss of 1.6 million disability-adjusted life years (DALYs) per year. This translates into a per capita loss of one to two days per year, and it is claimed that only air pollution imposes a larger disease burden.

The most widespread adverse effects of traffic noise are sleep disturbance and annoyance, which together constitute 92% of the loss (56% and 36%, respectively). The dominant noise source is road traffic, which subjects 54% of the EU population living in large urban areas to an average estimated noise exposure of Lden ≥ 55 dB outside their homes. For rail and air traffic noise, the corresponding percentages are 8% and 4%, respectively.[2]

While 54% of the EU population is exposed to Lden ≥ 55 dB road traffic noise outside their homes, 14% of the EU population is exposed to more than 65 dB, approximately 20% to 55–59 dB, and 20% to 60–65 dB noise.[2] Therefore, for the EU population living in large urban areas, normal noise exposure is characterized by Lden values of approximately 60 dB, while noise exposure above 65 dB is not uncommon. In fact, noise exposure of 65–70 dB is typical of the most exposed facades near roads in urban areas. The WHO guideline values for community noise state that the outdoor daytime and evening noise level should not exceed 50 dB for moderate annoyance,[3] while the guidelines of many EU countries allow daytime values of approximately 55 dB. It is common for there to be a gap of approximately 10 dB between actual immission levels and the guideline levels in use.

In planning new infrastructure and dwellings, it is difficult to bridge the gap between actual exposures and the guideline values. Political credibility is also at stake. Demanding the fulfilment of guideline values for new dwellings or infrastructure on health grounds is problematic when we have limited ability to alleviate the noisy conditions existing today. Moreover, there is still a gap of at least another 5 dB between the guideline levels and those characterizing a good environment.[4]

Making sufficient improvement entails using all available tools. We should be prepared to mix classical and novel noise control engineering approaches and to exploit the benefits of soundscaping. Reducing noise emissions from road vehicles is difficult and time-consuming for engineering and political reasons (e.g., Kihlman[5]), which is why noise problems have to be addressed by measures on both the emission and immission sides. In addition, the long-term and large-scale perspectives on urban and rural planning must be improved to avoid unnecessarily damaging the sound environments and to enable cost-effective abatement measures. Acoustic environment planning should be included in the planning process at the earliest possible stage and should always be proactive; that is, measures favourable to a good acoustic environment should be implemented even though there might not yet be a noise problem.

1.2 PRINCIPLES OF NOISE REDUCTION

The level of noise reaching a recipient is determined by two main factors: the strength of the source and the path of propagation. To abate road and rail traffic noise in general, we need to address both factors, that is, on both the emission side (i.e., the source strength) and the immission side (i.e., noise reduction during propagation).

1.2.1 Source strength

First, we should reduce the noise emission as much as is feasible. For road vehicles, the noise sources are of two main types: (1) traction sources, originating from the tyre–road contact, and (2) propulsion sources, originating from the engine and transmission and, for combustion engines, also from the exhaust system, air intake, fans, generators, and other auxiliary systems. There are also aerodynamic noise sources, which become dominant only at higher driving speeds. The noise emissions of road vehicles can generally be seen as a problem for which three parties are responsible: the road vehicle industry, the tyre industry, and the road owners. Propulsion noise is linked mainly to the vehicle industry, whereas the traction noise is shared between all three responsible parties.

As a result of the European HARMONOISE (a harmonized European method to assess environmental noise) and IMAGINE projects (improved methods for the assessment of the generic impact of noise in the enviroment), source models for road vehicles have been developed. The original Harmonoise model[6] used throughout the HOSANNA research project describes an average vehicle of each type driving on a reference road surface (see Jonasson[7] for further information on the source model). As its default case for urban roads, HOSANNA used two lanes of traffic, each modelled with a set of sources above the lane centre line.

Figure 1.1 shows the geometry of the sources and indicates how close to the road noise abatement devices, including ground treatments, can be positioned. In the default urban case, we assume a vehicle-type distribution of 95% light

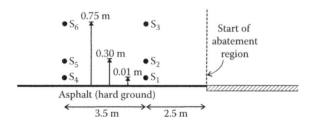

Figure 1.1 Positions of sources S1–S6 and the closest position of noise abatement treatments. Sources S1 and S4 are traction sources whereas the others are propulsion sources, the lower ones, S2 and S5, being for light vehicles and the upper ones, S3 and S6, for heavy vehicles. For noise exposure calculations, 20% of the source powers of traction and propulsion noise are exchanged in height, according to the original model.

vehicles and 5% trucks. The vehicle speed is 50 km/h and the flow is 20,000 vehicles per 24 hours. The default receiver heights used are 1.5 and 4 m.

Figure 1.2(a)–(d) shows the power spectra used for traction and propulsion sources at 50 km/h driving speed, as both A-weighted and linear (i.e., unweighted) levels. The maximum (unweighted) propulsion noise is obtained at lower frequencies, here at approximately 80 Hz, which is

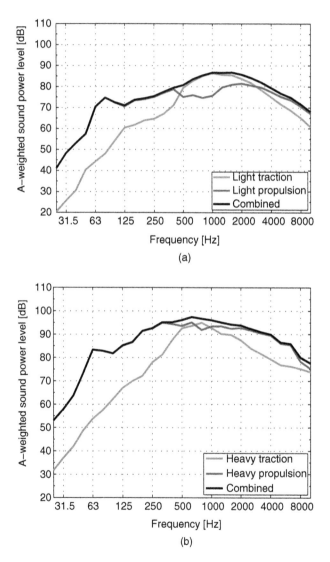

Figure 1.2 Spectrum of noise sources for light vehicles (a) and heavy vehicles (b) as A-weighted.

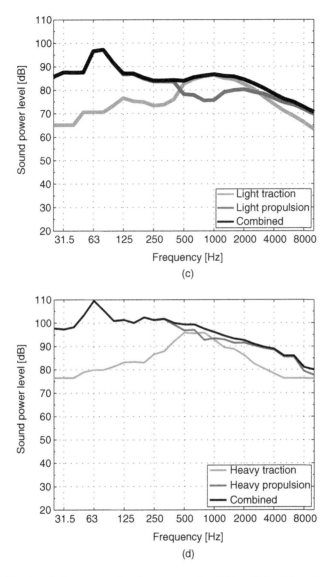

Figure 1.2 (Continued) Spectrum of noise sources for light vehicles (c) and unweighted (d) levels at a driving speed of 50 km/h.

related to the engine speed, or ignition rate. The dominance of propulsion noise at lower frequencies is persistent up to approximately 400 Hz for light vehicles, whereas for heavy vehicles propulsion dominates over the entire frequency range except at approximately 800 Hz, at which traction is about equally strong. Considering the single number dB(A) level at

50 km/h, traction noise is dominant for light vehicles and propulsion noise is dominant for heavy vehicles.

For the strengths of the different noise sources as a function of driving speed, a more recent model was used, which was formulated in the European CNOSSOS-EU joint research initiative.[8] Figure 1.3(a)(b) shows the single

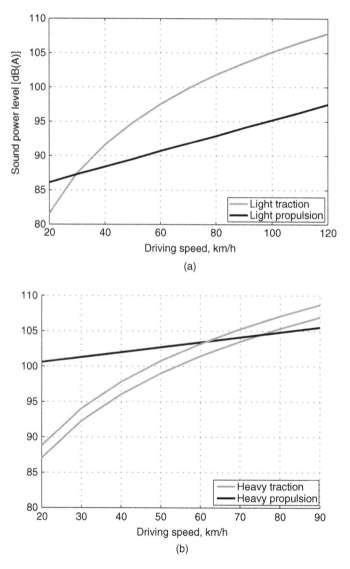

(a)

(b)

Figure 1.3 Speed dependence of traction and propulsion sources for light vehicles (a) and heavy vehicles (b). For the traction noise of heavy vehicles, two curves are displayed corresponding to four and six axles with tyres.

number dB(A) levels as a function of driving speed. For light vehicles, traction noise is seen to dominate above 30 km/h, whereas for heavy vehicles the transition occurs at approximately 70 km/h depending on the assumed number of wheel axles (the graphs shown in Figure 1.3(a)(b) are for four and six axles). It should be noted that these are the average results for the reference conditions. For example, accelerating vehicles or uphill driving leads to increased propulsion noise, whereas the use of road surfaces with smaller stone sizes or porous surfaces leads to reduced traction noise. In addition, individual driving styles and vehicles lead to different results. Due to the spectral and time pattern characteristics, propulsion sound may be audible even when traction sound dominates the A-weighted level. The performance of many noise abatement methods, including facade insulation, is normally better at higher than lower frequencies, meaning that the noise remaining after abatement is often more characterized by low-frequency sound (i.e., an increased proportion of propulsion noise sources).

Possible methods of abating noise emissions from propulsion sources include:

- Improved shielding of engine and power transmission (most relevant for heavy vehicles)
- Improved sealing of engine bay
- Reduced maximum engine power
- Improved mufflers to reduce noise from air intake and exhaust systems

Methods of abating traction noise originating from tyres include developing less noisy tyres for lower maximum speeds (e.g., 150 km/h) in a combined effort including low-rolling resistance, which would reduce fuel consumption. To abate traction noise originating from roads, measures involve quality control of newly laid surfaces, monitoring road pavement surfaces, optimising road texture, using smaller stones (especially relevant in Nordic countries), and using open porous asphalt.

For rail traffic, measures to abate propulsion noise include reducing noise from engines and fans as well as reducing aerodynamic noise (particularly for trains at higher speeds). Possible measures to abate propulsion noise originating from wheels and rails include improving brakes so as not to create deformed and noisy wheels (particularly in the case of freight trains), monitoring and maintaining rail surface quality, acoustically damping rails, ensuring good rail foundations, and preventing squealing by using a broad-enough curvature of the rail path as well as adequate friction (e.g., by using friction modifiers), particularly for tram and metro lines.

Politically determined limit values and appropriate test methods could drive the development of quieter vehicles. In addition, political instruments can influence the market, for example, encouraging the use of quieter vehicles and tyres. Furthermore, the purchase of vehicles (e.g., passenger cars, buses, trucks, and trams) by municipal and national governments could be more strongly guided by noise emission limit values.

1.2.2 Propagation effects

As sound propagates outdoors from a source, the factors determining the sound level at the receiver relate to the distance between source and receiver, the properties of the medium (air) in which sound propagates, and the properties of the boundary, i.e., the ground material and profile, including noise barriers and other obstacles.

1.2.2.1 Distance

In free space, sound from a point source spreads spherically and decays by 6 dB with each doubling of the distance from the source, whereas sound from a line source spreads cylindrically and decays by 3 dB per distance doubling. Predictions of the maximum sound levels of road traffic noise are based on a single vehicle as the noise source, whereas predictions of the average, or equivalent, sound level (e.g., Lden and LAeq, 24 h, in which time-averaged energy is used) assume the whole length of the road to be the source. Therefore, the maximum noise level decays by 6 dB per doubling of distance from the road, whereas the equivalent level decays by 3 dB per distance doubling, assuming a long, straight road and insignificant influence from the ground, wind, and other environmental factors (see Figure 1.4). A similar principle applies to railway noise, though multiple sources along the rail vehicle may contribute simultaneously to the maximum level. In general, when making noise-mapping calculations, the entire traffic network has to be considered along with the existing propagation conditions.

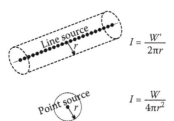

$$I = \frac{W'}{2\pi r}$$

$$I = \frac{W}{4\pi r^2}$$

Figure 1.4 For a line source (top), the sound intensity in the radial direction, I, decays inversely proportionally to the radial distance, r, whereas for a point source (bottom), the intensity decays inversely proportionally to the distance squared. Because the sound level varies as $10 \log_{10}(I)$, the two cases correspond to decays in level of 3 and 6 dB per distance doubling, respectively. The source strength is given as power per unit length, W' (in watts per meter), for the line source and as power, W (in watts), for the point source.

1.2.2.2 Medium

The acoustic properties of the noise propagation medium (air) relate to meteorological conditions, such as wind speed and temperature. The largest effects of such factors occur when they lead to refraction, i.e., the curving of sound paths. The degree of refraction is determined by the wind speed profile and the temperature variation with height, and the effect usually increases with propagation distance. As a result of downward curving, which may occur in the case of downwind sound propagation (i.e., wind blowing from the direction of the sound source and toward the receiver) or temperature inversion (i.e., increasing temperature with height), the noise levels may increase substantially. Conversely, upward curving, for example, under headwind conditions, may greatly reduce levels compared with situations without such refraction. Figure 1.5(a)–(d) shows the curving of the ray paths emanating from a point source under four types of refractive conditions: no refraction, wind, negative temperature gradient, and positive temperature gradient. These plots assume a linearly varying sound speed with height (i.e., a constant gradient), which causes the rays to be arcs of circles. Even though the circular arc ray model is approximate, it can be

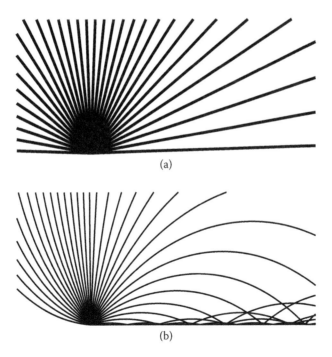

(a)

(b)

Figure 1.5 Ray paths for (a) a homogeneous atmosphere, (b) wind blowing from left to right.

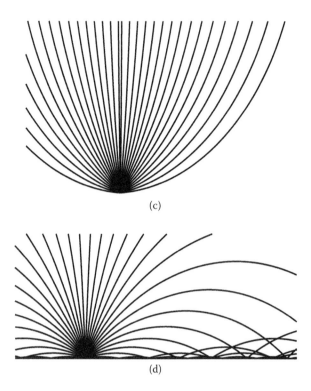

(c)

(d)

Figure 1.5 (Continued) Ray paths for (c) upward refraction due to decreasing temperature with height, and (d) downward refraction due to increasing
temperature with height.

used to point out the most relevant basic phenomena. Without refraction
(Figure 1.5(a)), the density of rays is independent of direction, and the
decreasing density with radial distance from the source corresponds to
the decay with distance. With upward refraction (left side Figure 1.5(b)
and both sides of Figure 1.5(c)), a shadow region is created into which no
sound ray enters, whereas with downward refraction, the sound is focussed
along the ground, where multiple reflections on the ground surface occur at
greater distances (right side of Figure 1.5(b) and both sides of Figure 1.5(d)).
It should be noted that both these effects are only approximately modelled
using ray theory. Ray theory relies on a high-frequency approximation,
whereby the effects of the medium and boundary at some distance away
from each ray are not considered, but may be treated in an extended model.
For example, a piecewise changing ground impedance can be modelled
using the Fresnel zone concept, and the nonzero sound field in a refractive
shadow can be modelled as diffraction plus a contribution from turbulence
scattering (e.g., References 9 and 10).

A better wind speed model is a logarithmic profile. The effect on the sound field of a logarithmic wind profile is illustrated in Figure 1.6 for (a) no refraction, (b) upwind sound propagation, and (c) downwind sound propagation. It can be seen that the upward refraction, under upwind conditions (b), leads to the formation of a shadow region near the ground at greater distances. The downward refraction, under downwind conditions (c), is seen to cause an increased sound level along the ground surface.

In the engineering prediction model resulting from the HARMONOISE and IMAGINE projects, combinations of linear and logarithmic sound

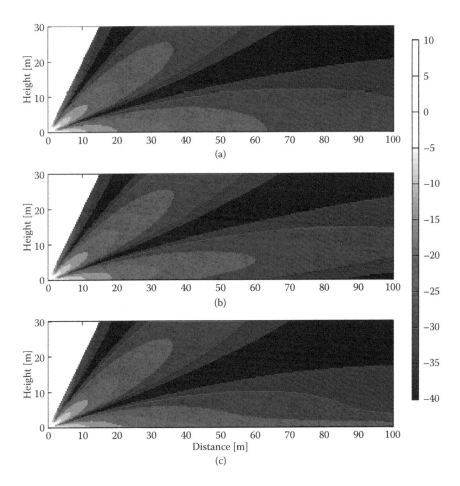

Figure 1.6 Sound field calculated using the parabolic equation method[11] at a single sound frequency of 500 Hz for (a) a homogeneous atmosphere, (b) a negative gradient, and (c) a positive gradient. The calculations are conducted assuming hard ground and a point source at a height of 0.5 m. The colour corresponds to the sound level relative to the free field level 1 m from the sound source.

speed profiles are used to model different meteorological situations, divided into five wind speed classes and five stability classes (determined by cloud cover and time of day).[6] As a result of that project work, it can be concluded that refraction may significantly affect traffic noise even at medium propagation ranges.[12] At a distance of 40 m from a motorway, the predicted effect is negligible, whereas at a distance of 160 m, the A-weighted sound level was predicted to increase by approximately 2–3 dB(A) under downwind conditions, and to decrease by at most 9 dB(A) under upwind conditions, for wind speed class 3 (3–6 m/s wind at a height of 10 m), with reference to wind speed class 1 (0–1 m/s wind at a height of 10 m), and stability class 3 (under cloudy daytime conditions). At a distance of 320 m, an increase of at most 7 dB(A) and a decrease of at most 13 dB(A) were predicted using the same reference conditions. The results from a corresponding measurement campaign were used to validate the model development, indicating good overall agreement (e.g., Reference 13).

Temperature, humidity, and, to a smaller extent, static pressure influence the degree of air attenuation, i.e., the molecular absorption of sound during its propagation. Air attenuation effects are of importance mainly at high frequencies. Atmospheric turbulence, in the form of random fluctuations in wind velocity and temperature, distorts the sound waves. The effects can be seen as the scattering of sound into shadow regions and the reduced strength of both positive and negative interference. These effects are important mainly at high frequencies.

Several of the following chapters discuss abatement methods that act on the medium of sound propagation, for example, planting trees behind barriers to improve barrier performance in downward refraction situations, planting a noise-reducing strip of trees along a road, or artificially creating upward refraction using sonic crystal-like barriers (see Chapters 2, 4, and 5[14–16]).

1.2.2.3 Boundary

In flat terrain, both direct sound from the source and ground-reflected sound can reach the receiver. The effect of the interaction between direct and reflected sound is called the ground effect. At some frequencies, direct and reflected sound partly cancel each other out, which causes the sound level to be lower than if the ground were not present. At other frequencies, the two sound waves reinforce each other, making the level higher than it would be if the ground were not present. For traffic noise propagating above an acoustically hard ground, such as asphalt, the two sound waves added together will normally lead to an increased noise level. However, above an acoustically soft ground, such as a lawn, the two waves may cancel each other out over a relatively broad frequency range, resulting in a lower noise level than if no ground were present.

At higher frequencies, the coherence between direct and reflected sound normally declines toward a purely energetic summation of the two contributions, e.g., due to the effects of turbulence or of random ground roughness.

Figure 1.7 illustrates the ground effect for a grass-type ground and a hard ground (e.g., dense asphalt surface) for a point source at a height of 0.75 m and a receiver at a height of 1.5 m located 50 m away. The result is shown both for a homogeneous atmosphere and with an estimated loss in coherence due to turbulence. The sound pressure level is plotted relative to free field, i.e., compared with a situation that has the same source and receiver positions but without a ground surface. As can be seen in the figure, the ground causes an increase of approximately 6 dB at the interference maximum, which is the expected pressure doubling for flat geometries (i.e., the source and receiver heights are small compared with the propagation distance). At the low-frequency limit, both ground types give rise to an interference maximum. The first interference dip for the hard ground is obtained near 4,000 Hz, whereas the first interference dip for the grass ground is at a lower frequency, 500–1,000 Hz, and is much broader. This difference in behaviour may cause a significant change in

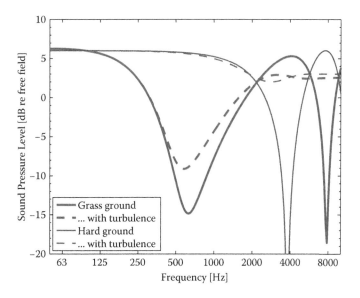

Figure 1.7 Ground effect plotted as sound pressure relative to free field as a function of frequency, for a grass-type ground and a hard ground; a point source is located at a height of 0.75 m and a receiver at a height of 1.5 m at a distance of 50 m. The acoustic impedance model used for the grass ground is according to the two-parameter model for hard, worn lawn,[17] from which the spherical reflection factor is calculated.

the A-weighted level of road and rail noise, even when including realistic reductions in coherence. The effect of turbulence is seen to limit the depths of the dips and the heights of the peaks by an amount that increases with frequency, eventually causing zero coherence between direct and reflected sound and a purely energetic summation at the high-frequency limit. When modelling an actual traffic situation using multiple sound sources that contribute simultaneously, the interference pattern will be more smeared and, thereby, closer to reality.

For conventional noise barriers, the height is the primary property, assuming the sound transmission through or around the sides of the barrier is negligible. To illustrate general barrier performance, the insertion loss of a thin, hard screen in the absence of a ground surface is plotted in Figures 1.8(a)(b) as a function of frequency, for a varying screen height

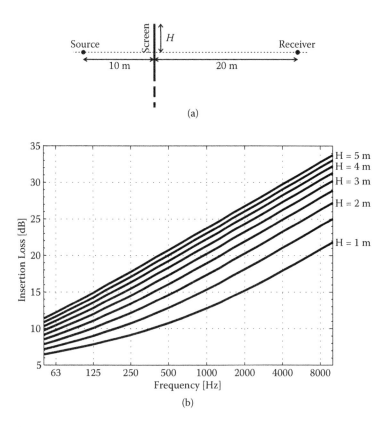

Figure 1.8 Acoustic insertion loss of a thin, hard screen of semi-infinite extension (ground surface is not included). Geometry (a); insertion loss in dB (b) for varying screen heights, *H*, above the line of sight, of 1–5 m in 0.5-m steps. (Diffraction model according to Pierce.[18])

and otherwise fixed locations of source, screen, and receiver. For a flat and acoustically hard ground surface directly below source and receiver, the insertion loss would decrease by approximately 6 dB. (For taller screens, the decrease tends to be 6 dB, whereas for lower screens the multiple reflections between screen edge and ground alter the behaviour.) The insertion loss is low at low frequencies and increases with frequency. At higher frequencies, it tends to increase by 3 dB per frequency doubling. At lower frequencies, the dependence on screen height is weaker than at higher frequencies. In real situations, the insertion loss can be reduced depending on meteorological conditions, mainly downward refraction under downwind propagation conditions and scattering by turbulence at higher frequencies.

If the top of the barrier is widened, the acoustic effect will be further improved. Concerning the barrier location, better performance is generally achieved if the barrier is placed near the source or near the receiver. In an inner city environment, it may be preferable, therefore, to use noise barriers of a relatively low height if they can be located near the traffic sources. To further improve the performance of such barriers, their width can be increased and the materials on their tops and faces should be carefully chosen. The materials should be acoustically soft (e.g., porous), and in urban environments characterised by multiple sound reflections, it is especially important to choose acoustically absorbent materials. In general, significant reflection can occur from the facades in street canyons, from the faces of noise barriers, and in vehicle bodies, particularly for large, heavy road and rail vehicles.

In general, urban planning greatly affects noise exposure by addressing the placement and regulation of surface transport in connection with building form and function. As described in Chapter 7, roof and facade treatments are used to reduce the noise level in shielded areas, such as inner yards, which may lead to annoyance reduction by applying the "quiet side" concept to dwellings.

Attenborough et al.[19,20] discuss several methods that act on the boundary, including softening hard ground, roughening flat ground, appropriate barrier design, and using acoustically absorbent materials, such as vegetation substrate, on barriers, facades, and roofs (see Chapters 6 and 7).

In noise-mapping calculations, all the above propagation effects could be considered, and the effects of noise barriers, ground type, and terrain profile could be estimated. Furthermore, using local meteorological statistics allows one to estimate the distribution of noise immission levels due to various meteorological conditions. However, the simplifications made in current noise-mapping methods mean that the acoustic effects of many of the abatement tools presented here are not evident; accordingly, improved physical modelling would permit the further development of noise-mapping methods.

1.2.3 Noise indicators

To estimate the effects on the health and well-being of various abatement measures, we need to keep in mind that single-value noise levels are indicators with limited precision. Two different sound environments, e.g., hearing a distant motorway or a nearby urban road, may correspond to the same Lden value. In addition, even for the same sound, the response varies between individuals. Different temporal patterns play a role as do the spectral characteristics, e.g., the low-frequency character in the shadow of a noise barrier compared with the brighter direct sound of a nearby vehicle. When examining perceived annoyance, it is of interest to consider these aspects together with background sounds, which may even be positive, as in the case of sounds from fountains and singing birds. The temporal character also strongly influences sleep disturbance, which is strongly related to the number of events (e.g., car pass-bys).[21]

The spectral effects of various abatement measures also must be considered. In the case of facade insulation, for example, the noise blocked by a barrier will not be reduced by as many dB(A) as will an unblocked noise transmitted through a window, due to the low-frequency shift caused by the barrier. If many abatement tools are used at the same time, their individual single number effects in dB(A) cannot always be summed to predict the total effect. Additivity holds strictly only when the individual effects are the same at all frequencies. It is usually easier to reduce the higher frequencies in noise using passive methods for absorption or reducing transmission. To abate the lower-frequency components, larger devices are usually needed, or one might use resonant absorbers or even active elements, such as loudspeakers.

1.3 CONCLUDING REMARKS

The general principles of outdoor acoustics have been briefly described together with the basics of noise reduction, including the key link between noise indicators and perception. The source strengths (i.e., emissions) and their reduction have been characterized, focussing on the dominant road traffic noise. It is stressed that the simplifications made in current noise mapping methods lead to insensitivity to many of the noise-reducing tools suggested in the HOSANNA research project. It is essential, in view of the long-term aim of reducing noise impact in Europe, to improve physical modelling in order to develop better noise-mapping methods. Finally, it is important to be aware that many of the noise-reducing tools suggested in HOSANNA need careful consideration very early in the planning processes if they are to be cost-effectively implemented.

REFERENCES

1. WHO. 2011. *Burden of disease from environmental noise.* Copenhagen: World Health Organization Regional Office for Europe.
2. European Environment Agency (EAA). 2013. *NOISE—Noise observation and information service for Europe.* Online at: www.eea.europa.eu/data-and-maps/explore-interactive-maps/promotions/noise-2014-noise-observation-and-information-service-for-europe (retrieved December 11, 2013).
3. WHO. 2000. *Guidelines for community noise.* Geneva: World Health Organization.
4. Kihlman, T., and M. E. Nilsson. 2010. 40-45 dB—Criteria for a good sound environment in urban areas. Paper presented at the *Proceedings of Inter-Noise 2010*, INCE, Lisbon.
5. Kihlman, T. 2006. Sustainable development in an urbanizing world: The noise issue. *Noise News International* 14:14–19.
6. Nota, R., R. Barelds, and D. Van Maercke. 2005. Harmonoise WP 3 engineering method for road traffic and railway noise after validation and fine tuning. Deliverable of WP3 of the HARMONOISE project. Document ID HAR32TR-040922-DGMR20.
7. Jonasson, H. 2007. Acoustical source modelling of road vehicles. *Acta Acustica United with Acustica* 93(2):173–184.
8. Kephalopoulos, S., M. Paviotti, and F. Anfosso-Lédée. 2012. *Common noise assessment methods in Europe (CNOSSOS-EU).* Luxembourg: Publications Office of the European Union.
9. Plovsing, B. 2007. *Proposal for Nordtest method: Nord2000—Prediction of outdoor sound propagation.* Report AV 1106 (07). Copenhagen: DELTA Acoustics.
10. Salomons, E., D. van Maercke, J. Defrance, and F. de Roo. 2011. The Harmonoise sound propagation model. *Acta Acustica United with Acustica* 97(1):62–74.
11. Forssén, J. 1998. Calculation of sound reduction by a screen in a turbulent atmosphere using the parabolic equation method. *Acta Acustica united with Acustica* 84(4):599–606.
12. HARMONOISE. 2005. *Validation of the Harmonoise models.* Harmonoise Technical Report no. HAR28TR-041109-TNO11. TNO-TPD, The Netherlands.
13. van Maercke, D., and J. Defrance. 2007. Development of an analytical model for outdoor sound propagation within the Harmonoise project. *Acta Acustica United with Acustica* 93(2):201–212.
14. Defrance, J., P. Jean, F. Koussa, T. Van Renterghem, J. Kang, and J. Smyrnowa. 2014. Innovative barriers. In *Environmental methods for transport noise reduction*, eds. M. E. Nilsson, R. Klæboe, and J. Bengtsson (pp. 19–47). Oxford, UK: Spon Press.
15. Van Renterghem, T., D. Botteldooren, J. Kang, and H.-S. Yang. 2014. Acoustical characteristics of trees, shrubs and hedges. In *Environmental methods for transport noise reduction*, eds. M. E. Nilsson, R. Klæboe, and J. Bengtsson. Oxford, UK: Spon Press.
16. Van Renterghem, T., K. Attenborough, and P. Jean. 2014. Designing vegetation and tree belts along roads. In *Environmental methods for transport noise reduction*, eds. M. E. Nilsson, R. Klæboe, and J. Bengtsson. Oxford, UK: Spon Press.

17. Attenborough, K. 1992. Ground parameter information for propagation modeling. *The Journal of the Acoustical Society of America* 92:418.
18. Pierce, A. D. 1974. Diffraction of sound around corners and over wide barriers. *The Journal of the Acoustical Society of America* 55:941.
19. Attenborough, K., S. Taherzadeh, I. Bashir, J. Forssén, B. Van der Aa, and M. Männel. 2014. Porous ground, crops and buried resonators. In *Environmental methods for transport noise reduction*, eds. M. E. Nilsson, R. Klæboe, and J. Bengtsson. Oxford, UK: Spon Press.
20. Attenborough, K., I. Bashir, T. J. Hill, S. Taherzadeh, J. Defrance, and P. Jean. 2014. Noise reduction using surface roughness. In *Environmental methods for transport noise reduction*, eds. M. E. Nilsson, R. Klæboe, and J. Bengtsson. Oxford, UK: Spon Press.
21. WHO. 2009. *Night noise guidelines for Europe*. Copenhagen: World Health Organization Regional Office for Europe.

Chapter 2

Innovative barriers

Jérôme Defrance, Philippe Jean, Faouzi Koussa,
Timothy Van Renterghem, Jian Kang, and Yuliya Smyrnova

Abstract. Innovative barriers using natural materials have been shown to be quite effective, promising solutions to significantly abate ground transportation noise in urban and rural areas. In this chapter, the acoustic performance of a large set of such novel solutions has been numerically evaluated in various environmental configurations and types of transportation infrastructures: urban streets, tramways, motorways, and railways. For urban streets, low-height vegetated and gabion (cage filled with earth or rocks) barriers (no more than 1 m high) are studied. It is shown that the noise reduction is 5 dB(A) maximum for a four-lane street. If extra interlane low barriers are added, the abatement may reach 10 dB(A) up to 5 m high (more when closer to the ground). For urban tramways, low-height berms and sonic crystal-assisted barriers are proposed and assessed. The addition of an extra intertrack absorbing low barrier, as well as the addition of a low vegetated barrier at the edge of a bridge over a pedestrian path, also are addressed. The predicted insertion loss is large, up to 15 dB(A), because the innovative barrier can be built very close to the noise sources. In rural situations (motorways and railways), the acoustic attenuation due to the addition of a row of trees behind a conventional 4-m-high barrier, the addition of a vegetated cap atop the barrier, and the addition of nonflatness on conventional 4-m-high berms have been assessed. The extra acoustic gain ranges from 3 dB(A) (highest receivers) to 8-dB(A) (lowest ones).

2.1 INTRODUCTION

The conclusions and recommendations presented in this chapter are related to work on innovative barriers carried out in the research project HOSANNA.[1–4] Each section of this chapter refers to a specific family of ground transportation infrastructure: urban streets (Section 2.2), tramways (Section 2.3), motorways (Section 2.4), and railways (Section 2.5).

2.1.1 Receiver zones

Most noise propagation calculations were carried out for four different 20-m-long, 1-m-high receiver zones as shown in Figure 2.1. The lower zones

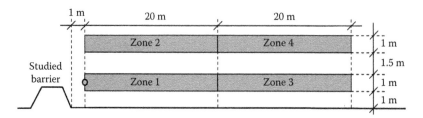

Figure 2.1 Definition of the four receivers' zones and the "pedestrian" receiver (circle).

1 and 3 (extending from 1 to 2 m in height) are located at heights around 1.50 m above ground (pedestrian, cyclist, or building ground floors). The upper zones 2 and 4 (extending from 3.50 to 4.50 m in height) are located at heights close to 4 m above ground (first floor of buildings).

We also considered noise affecting a single receiver located 1.50 m high, 1 m away from the studied noise protection. This receiver point represents the location of a nearby pedestrian or cyclist and, hereafter, is called *pedestrian*.

2.1.2 Objectives

The results of the simulations carried out were evaluated with respect to the following quantitative objectives as measured at a 4-m-high receiver location alongside a given surface transport corridor, compared to an untreated situation:

- To produce designs for vegetated low barriers (or low barriers using natural materials), which leads to a minimum noise abatement of 6 dB(A) in urban areas and 3 dB(A) in rural areas, at a 4-m-high receiver location alongside a given surface transport corridor, compared to an untreated situation.
- To produce designs for combination of innovative barriers and shaped soil, which leads to a minimum noise abatement of 3 dB(A) at a 4-m-high receiver location alongside a given surface transport corridor, compared to a situation with a straight barrier alone.
- To produce designs for combination of innovative barriers and vegetation (on barriers' tops, trees behind barrier, etc.) that leads to a minimum noise abatement of 3 dB(A) at a 4-m-high receiver location alongside a given surface transport corridor, compared to a situation with a straight barrier alone.

All the above values are increased by 2 dB(A) for receiver positions at a 1.5-m height. The objectives are listed in Table 2.1.

Table 2.1 Noise reduction objectives

Noise mitigation method	Noise reduction in dB(A)				Reduction in relation to:
	Urban environment		Rural environment		
	4 m[a]	1.5 m[a]	4 m[a]	1.5 m[a]	
Low barrier	6	8	3	5	No treatment
Low berm	6	8	3	5	No treatment
Substrate cover			3	5	4-m-high rigid barrier
Vegetated caps			3	5	4-m-high rigid barrier
Berms			3	5	4-m-high rigid barrier

[a] Receiver height.

2.2 URBAN STREETS

2.2.1 Building low-height vegetated barriers

Different combinations of small barriers (height 0.96 m, width 0.64 m) in a typical urban street were considered, including a single barrier at road side or middle of road (L/R or M, see Figure 2.2), two barriers (L + R), or three barriers (L + M + R). The traffic characteristics were as follows: 95% light vehicles and 5% heavy vehicles, with results averaged for speeds between 30 and 70 km/h. Reflections between opposite facades lead to a strong increase in sound pressure levels of road traffic noise. The realism of the simulations was increased by considering four traffic lanes and three source heights following the Harmonoise/Imagine road traffic source power models,[5] and by using profiled facades inducing some diffuse scattering that might affect the efficiency of the low barriers.

Because this case was based only on numerical calculations, a validation point has been defined for detailed comparison between various full-wave techniques. Very good agreement was obtained in this cross-validation exercise.

The barriers on both sides of the four lanes and on the middle reservation were considered to be vegetated with the typical substrates used for wall vegetation because they show good acoustic properties. The measured material properties of these were used in the calculations. Geometrical details of the street canyon are found in Figure 2.2.

2.2.1.1 Conclusions

Although the sources and receivers are located in the same reverberant space (namely, the street canyon), noise reduction can be achieved using low-height barriers. Effects are most pronounced along the walkways: up to 5 dB(A) total traffic noise insertion loss was predicted in the configuration in Figure 2.2.

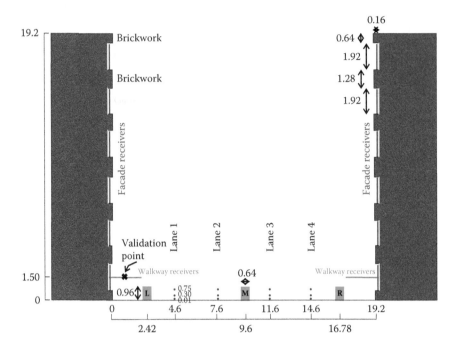

Figure 2.2 Numerical study setup for low-height screens in a street canyon.

Along the facades, some shielding is possible, but, of course, it is limited because a large part of the facades receive a direct sound contribution. Absorption at the surfaces of the low-height noise barriers is an important prerequisite to seeing significant effects in canyon streets. The more absorbing the barrier; the higher the effect for multilane road traffic noise reduction. The increase in insertion loss from configuration L + R with barriers on both sides of the four lanes to L + M + R, where a barrier is added to the central reservation, is only moderate for walkway receivers. In contrast, the increase from a single low-height barrier (configuration M) to a two-barrier configuration (L + R) is large.

Additional calculations with a full 3D model verified that 2D approaches are accurate when considering spatially averaged total traffic noise insertion losses.

2.2.1.2 Recommendations

In a nonreverberant environment, low-height barriers should be placed close to the source because it is only under this condition that rolling noise engine noise can be shielded.

In reverberant environments, such as street canyons, the main effect is not so much the shielding provided by the small barriers, but rather the addition of absorption. The surface properties of these barriers are then very important. Calculations showed that in the case of fully rigid screens (case L + M + R), only 1 dB(A) insertion loss is obtained. The use of the vegetated wall substrate enables the low barriers to absorb sound at low frequencies reasonably well. Because vehicle speeds are rather low in urban streets, with a significant part of the noise energy from vehicle propulsion in the lower frequencies, adding this type of substrate to the barriers is of special interest.

As an alternative to only using absorbing barriers at street level, absorption may be applied to facades as well. To reduce sound pressure levels in the street, the lower part of the facade should be the main point of focus. The predicted insertion losses are rather modest, but the method still may be considered given the lack of other noise abatement solutions for such cases.

2.2.2 Adding low-height vegetated interlane barriers

The aim was to assess the shielding effect of adding a low absorbing barrier between two traffic lanes when a low barrier was already built at the edge of the infrastructure, along the walkway. Calculations were made for 1-m-high barriers made of vegetation substrate (40 cm thick) with a thin rigid core. A four-lane urban street was considered (B and D being the interlane barriers; see Figure 2.3), with 95% light and 5% heavy vehicles travelling at 50 km/h. A 50-cm area on both sides of the low barriers; was reserved for emergency purposes resulting in a minimum distance between vehicles and barrier of 0.5 m. Three situations were simulated: open area (no building), building on receiver side, and buildings on both sides.

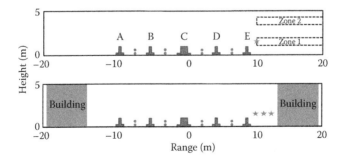

Figure 2.3 Pedestrian receiver locations (*) and possible locations of the low barriers (without or with buildings): A-B-C-D-E, B, and D being the interlane barriers.

2.2.2.1 Conclusions

The extra noise attenuation due to the addition of a central interlane, low-absorbing barrier between the two first lanes (barrier D in Figure 2.3), between the two right lanes of a four-lane urban street already protected with an edge barrier on the right side (barrier E in Figure 2.3) in an open area (no buildings), is 10 dB(A) for the pedestrian receiver. The noise reduction is 5 dB(A) in the 1.5-m-high receiver zones and 2–3 dB(A) for the 4-m-high receiver zones. In built-up areas (a building on receiver side or buildings on both sides), the extra noise attenuation provided by the interlane barrier D is in the range 4–5 dB(A) for the three 1.5-m-high "pedestrian" receivers.

Given a configuration with edge barriers, the addition of a central low-absorbing barrier at the very middle of the road infrastructure (barrier C in Figure 2.3) results in an extra noise attenuation of 3–5 dB(A). These values also apply to situations where a preexisting concrete (e.g., security) barrier is covered with absorbing vegetation substrate.

If noise abatement is sought on both sides of the four-lane street, an extra interlane, low-absorbing barrier also has to be installed between the two lanes that are farthest away (barrier B in Figure 2.3).

2.2.2.2 Recommendations

The shielding effect of adding a low-absorbing barrier between two traffic lanes can be important. We recommend using this principle for creating noise shielded areas close to road infrastructures, which may be desirable when a cycle path is located along the road or when the road runs along a square, an urban park, or a recreational area.

This principle could be applied as well in sufficiently wide canyon streets (such as avenues and boulevards) where the ambient noise for pedestrians walking on the pavement or cyclists moving along the road is often too high. This type of action would improve the environmental quality of such places (shopping entrances, outdoor cafés, and touristic walkways).

The shielding effect is also efficient for the lowest floors (up to 5 m high) of buildings built along urban roads. However, for higher floors, such absorbing low barriers have only small noise abatement impact.

2.2.3 Building low-height gabion barriers

The aim was to assess the shielding effect of 1 m × 1 m noise barriers made of stones (gabions) in an open area. The simulations were carried out in the case of a two-lane urban road (Figure 2.4) with 100% light vehicles (50 km/h). Four types of gabions using two different stone sizes were simulated (Figure 2.5).

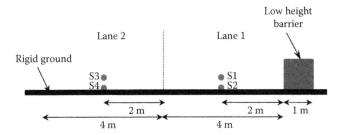

Figure 2.4 Urban road configuration with gabions.

Figure 2.5 Geometry of studied gabions: homogeneous (GBI and GB2) and layered structures (GB3 and GB4).

2.2.3.1 Conclusions

The extra noise attenuation due to the addition of 1 m × 1 m gabion barrier (in comparison with a situation without a barrier) along a two-lane urban road is in the range 7–8 dB(A) for pedestrians and bicyclists (1.5-m-high receiver zones) and 5–6 dB(A) for the first floor, 4-m-high receiver zone. The largest noise reduction is obtained with gabion cross-sectional geometry GB3 (Figure 2.5), but the acoustical performances of all tested gabions were of the same order (±1 dB(A)).

2.2.3.2 Recommendations

The use of gabion barriers along one- or two-lane urban roads may be recommended and they represent easy-to-implement measures.

This solution is efficient for low receivers that are in the shielded zone, not too close to the barrier. Therefore, we recommend preserving a space of at least 1 m between the gabion barrier and the walkway or cycling path.

2.3 TRAMWAYS

2.3.1 Building low-height earth berms

The aim was to assess the shielding effect of low-height grassy earth berms in open areas (no buildings nearby) whose height is 1 m and

Figure 2.6 Geometries of the studied 1-m-high earth berms (B1 to B6).

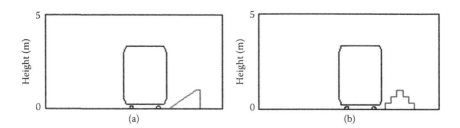

Figure 2.7 (a) General overview for berms B2 and (b) B5 for the tramway configuration (closest tram).

cross-sectional area is 1 m². The geometries of the studied berms are illustrated in Figure 2.6, and the two-track tramway (running at 30 km/h) in Figure 2.7(a)(b).

2.3.1.1 Conclusions

The extra attenuation due to the addition of a low earth berm along a two-track tramway on flat terrain is in the range of 9–12 dB(A) for pedestrians and bicyclists (1.5-m receiver zones) and 7–12 dB(A) for the first floors of roadside buildings (4-m-high receiver zone). The highest average extra attenuations are obtained for the "stair shapes" B5 and B6, which are defined in Figure 2.6.

2.3.1.2 Recommendations

Low-height earth berms proved to be a possible noise-shielding solution for receivers on the first floors of nearby buildings (up to 5 m high). In the case of flat terrain, we recommend their use when they can be located close enough to the noise source, e.g., along tramways.

We recommend "square" or "stairs" low-earth berms (shapes B1, B5, and B6 in Figure 2.6) instead of the more conventional triangular shapes (B2, B3, and B4).

2.3.2 Building low-height, sonic crystal-assisted barriers

The aim was to assess in an open area the shielding effect of 1 m × 1 m noise barriers made of spaced horizontal wooden cylinders (sonic crystals). The simulations are carried out for the case of a two-track tramway (Figure 2.8) with trams moving at 30 km/h. Two types of innovative barriers are studied: with and without 1-m-high straight, rigid barriers at the back (Figure 2.9(a)(b)). The ground is considered to be hard.

2.3.2.1 Conclusions

The extra noise attenuation due to the addition of a 1 m × 1 m sonic crystal-assisted barrier along tramways, for all receiver zones, is in the range

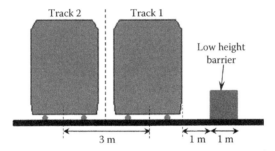

Figure 2.8 Tramway configuration with sonic crystal-assisted barriers.

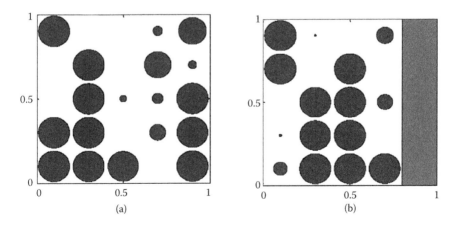

Figure 2.9 Example of sonic crystal arrangements (sources on the left side, receivers on the right side). Wood cylinders: alone (a) or assisting a low rigid, straight barrier (b).

5–6 dB(A) for a sonic crystal system alone, and 8–9 dB(A) when coupled with a 1-m-high rigid, straight barrier. Complementary calculations have shown that, in the latter case when covering the cylinder surface with an absorbing material, the extra noise attenuation may be up to 15 dB(A). This high value is mainly due to the large decrease by absorption of the multiple sound reflections between the tram body and the barrier (so-called *body-barrier effect*).

2.3.2.2 Recommendations

Solutions based on sonic crystals along tramways may not be easy to construct and design. On the other hand, using such noise systems may be motivated for aesthetic reasons.

This solution is efficient for low receivers only when they are in the shielded area not too close to the barrier. Therefore, we recommend preserving a space of at least 1 m between the sonic crystal-assisted barrier and the walkway or cycling path.

2.3.3 Adding low-height vegetated intertrack barriers

The aim was to assess the shielding effect of adding a low-absorbing barrier between two tramway tracks when a low barrier is already built at the edge of the infrastructure, along the walkway. Predictions were conducted for 1-m-high barriers made of vegetation substrate (40 cm thick) with a thin rigid core. The configuration was a two-track tramway with trams moving at 30 km/h (A being the intertrack barrier; see Figure 2.10). Three urban situations were simulated: open area (no building), building on receiver side, and buildings on both sides.

2.3.3.1 Conclusions

The extra noise attenuation due to the addition of a central intertrack low-absorbing barrier on a two-track tramway in an open area (no building) is in the range 6–8 dB(A) for all receiver zones (see Figure 2.10). A value of 9 dB(A) is obtained for the "pedestrian" receiver. These results remain of the same order in built-up areas for the three 1.5-m-high "pedestrian" receivers (see Figure 2.1): 8 dB(A) in the case of one building on receiver side and 7 dB(A) in the case of buildings on both sides.

2.3.3.2 Recommendations

The shielding effect of adding a low-absorbing barrier between two tramway tracks can be important. We recommend that this principle is used to

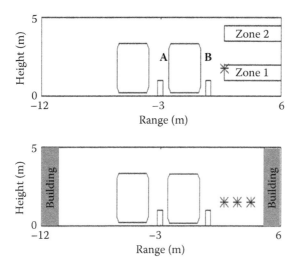

Figure 2.10 Pedestrian receivers locations (*) and possible locations of the low barriers (without or with buildings): A–B, A being the intertrack barrier.

create noise-shielded areas close to tramway infrastructures, e.g., when a cycle path is constructed along tramway tracks or when tracks run along a square, an urban park, or a recreational area.

This principle could be applied also in sufficiently wide canyon streets (such as avenues and boulevards) where the ambient noise for pedestrians walking on the pavement or cyclists moving along the road is often too high. This type of action would improve the quality of such places.

The shielding effect is also efficient for the lowest floors (up to 5 m high) of buildings built along roads and tramways. However, for higher floors, such absorbing low barriers have only small noise abatement impact.

2.3.4 Building low-height vegetated barriers at the edges of bridges

In this section, we assess the extra noise attenuation provided by a 1-m-high absorbing barrier (consisting of a 40-cm-wide vegetation substrate surrounding a thin rigid core), compared with a nonabsorbing barrier built at the edge of a 6-m-high bridge (on one or two sides) in the case of a two-track tramway (trams moving at 30 km/h), as shown in Figure 2.11.

We focus here on the pedestrian on a walkway or the cyclist moving below the bridge, the receivers being at a height of 1.5 m (dots in Figure 2.11).

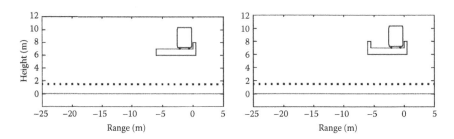

Figure 2.11 Overview of bridge with barrier on one or two sides: Tramway case.

2.3.4.1 Conclusions

The 40-cm-wide, 1-m-high barrier made from vegetation substrate over a core is very efficient for improving the noise environment below the bridge in areas designated for pedestrians and cyclists. Even without the inner rigid core, a high noise-reducing effect is predicted for both road and tramway infrastructure types.

The extra noise-reducing effect is about 12 dB(A), the attenuation being in the range 16–20 dB(A) for the closest tramway alone.

2.3.4.2 Recommendations

The shielding effect of adding a low-absorbing barrier at the edge of bridges can be large, for both road and tramway infrastructures. We recommend the use of low barriers to improve the noise environment below bridges, notably where pedestrian and cycle paths have been designed.

One may note that the addition of a 1-m-high absorbing barrier at the edge of bridges does not prevent tramway passengers from seeing the landscape beyond the bridge. The light weight of vegetation substrate also makes the product suitable for bridges.

By creating noise-shielded areas below bridges, we aim at promoting environmentally friendly transportation modes (cycling, walking) as well as urban parks and recreational areas along rivers where natural environmental characteristics have to be preserved.

2.4 MOTORWAYS

2.4.1 Covering conventional rigid noise barrier with vegetation substrate

We assessed the acoustical abatement due to covering a conventional 4-m-high rigid barrier with a layer of vegetation substrate for a four-lane motorway. The modelled traffic scenario was composed of 85% light

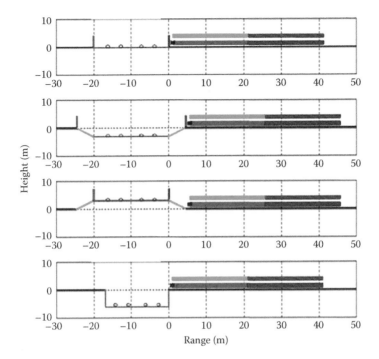

Figure 2.12 Geometry of four studied terrains (from top to bottom): flat, depressed, embanked, and trench. Colours show the four receiver zones and the circle is the pavement receiver position.

vehicles (120 km/h) and 15% heavy vehicles (90 km/h). Four configurations were studied: flat terrain, depressed road (3 m below the flat terrain), road on embankment (3 m above the flat terrain), and 17-m- and 20-m-wide trenches (6 m deep). (See Figure 2.12, where the vegetation substrate is applied to the side of the barriers/trenches facing the road.)

The vegetation substrate may partially (half surface) or totally cover the straight, rigid barriers (or walls, in the case of the trenches) on surfaces facing the road. In the case of partial covering, different arrangements were tested: lower half, middle half, upper half, and alternative strip of 0.5 m from bottom to top (rigid/absorbent/rigid, and so on). The calculations were carried out for the four zones and the "pedestrian" receiver position (detailed in Figure 2.1).

2.4.1.1 Conclusions

In the case of a flat terrain, the additional noise abatement effect obtained from covering the two barriers with the substrate is in the range 5–6 dB(A) for the lower receiver zones and 7–8 dB(A) for the upper ones. For the "pedestrian" receiver, the effect is 4 dB(A). In the case of an embankment and full covering, the noise-reducing effects are generally lower than those relative to flat terrain by a couple of dB; the effect of covering the two barriers is in the range 4–5 dB(A) for the lower receiver zones and 5–6 dB(A) for the upper ones. For the "pedestrian" receiver, the effect of adding absorbing surfaces is 2 dB(A).

In the case of depressed infrastructure (full covering), the effects of adding absorbing surfaces to the abatements are generally higher than those relative to flat terrain by a few dB; the effect of adding the absorbing substrate to the two barriers is in the range 6–7 dB(A) for the lower receiver zones and 7–8 dB(A) for the upper ones. For the "pedestrian" receiver, the effect is 5 dB(A).

When only half of the barrier's surface is absorbent, a "horizontal strips" arrangement (alternative succession of rigid and absorbent parts on the surface) shows the best performance, with an average loss in noise reduction of 1–3 dB(A), compared to the fully covered situation. This loss is generally greatest when the middle or upper half is covered (3–8 dB(A)).

In the case of trenches, the narrower the trench, the higher the acoustic gain. A full covering of the two walls gives an additional noise reduction in the range of 10–12 dB(A) for the lower zones and 7–12 dB(A) for the upper ones. For the "pedestrian" receiver, the additional noise reduction provided by covering the surfaces is only 2–3 dB(A) because some sources are in direct sight. When only half of the wall surface is absorbent, noise reduction is reduced by 3–4 dB(A) for all receiving zones. While the type of arrangement is not of major importance, the lower-half arrangement shows a better performance.

2.4.1.2 Recommendations

These results show that the noise abatement obtained with a vegetation substrate on rigid barriers is close to the one observed with conventional absorbent materials currently used for such conventional barriers and trench configurations (several meters high). Thus, vegetation substrate appears to be a very effective way of alternatively treating most conventional straight, rigid barriers and walls. The addition of living plants does not significantly affect the acoustic absorption of the layer of soil and can even enhance it, depending on the size of the leaves (see Horoshenkov et al.[6]; also Chapter 3: Acoustic Performance of Vegetation and Soil Substratum in an Urban Context).

This solution using soil substrate would increase vegetation along motorways and, therefore, would improve the global appearance of a conventional absorbing noise barrier, contribute to biodiversity, and help in avoiding graffiti, while providing equivalent acoustic performance. If the road is located in a trench or cutting, it has been shown that when the receiver is not too close to the upper ground edge, the noise reduction may be quite large. It also has been pointed out that covering the lower parts of the walls of the trench or cutting gives greater reductions than when covering the upper parts. Moreover, the presence of plants may visually contribute to a more pleasant ambiance and the substrate may help in air pollution reduction. So, this type of solution is highly recommended for such confined configurations.

2.4.2 Adding a row of trees behind a conventional noise barrier

The size of the protected area and the size of the noise reductions provided by noise barriers are significantly reduced under downwind conditions. The effect of rows of trees, acting as windbreaks under downwind conditions, was assessed for a typical traffic noise situation. On either side of the road, a straight, 4-m-high (rigid) noise barrier is present, bordering a four-lane road (Figure 2.13). The traffic consists of 100% light vehicles and vehicle speeds ranging between 30 and 130 km/h. A downwind receiver zone is defined. Due to the screen-induced refraction of sound by wind, downward refraction into the shadow zone is expected, limiting the barrier shielding. A significant improvement in the acoustical performance due to the presence of the trees is predicted. (See Van Renterghem et al.[7] and Chapter 5 (Designing Vegetation and Tree Belts along Roads) for a detailed discussion of the use of trees behind noise barriers to improve shielding effects.)

2.4.2.1 Conclusions

The maximum (adverse) wind effect appears at a greater distance for moderate winds compared to strong winds. Up to 10 dB(A) of the shielding can be lost in the case of highway noise. In such windy conditions, the noise may even exceed the noise level without wind on a completely unshielded highway; it is as though the barrier were absent. Dense canopies are useful

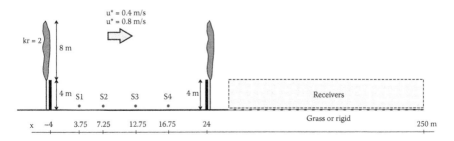

Figure 2.13 Setup for rows of trees behind 4-m-high straight barriers.

as windbreaks. Close to the barrier, an excellent improvement in shielding under wind is obtained; the negative action of the wind almost can be fully counteracted by the presence of the trees. Further analysis shows that low frequency sound is hardly affected downwardly by the action of wind in a shielded zone, while high frequency sound is strongly refracted. At frequencies above 1 kHz, no or even negative insertion loss (relative to unobstructed terrain in the absence of wind) is expected in strong winds (in the absence of trees). Thus, trees are especially interesting to increase shielding at higher sound frequencies in wind.

However, care is needed. The screen-induced vertical gradients in the horizontal component of the wind speed do not disappear. They are reduced to some extent, but also are shifted to other spots near the shielded zone behind the barriers. This has been clearly shown by analysis of these gradients. Typically, such gradients move to a greater height above the tree-tops. This means that the location where the sound energy bends toward the ground is shifted upward and away from the barrier. For receivers at short distances behind the barrier, this is clearly positive, because the sound energy passes overhead. However, receivers farther downwind from the barrier are now exposed to more of the sound energy, thus reducing the performance of the noise barrier at longer distances.

2.4.2.2 Recommendations

The use of a row of trees behind a conventional noise barrier is especially useful in a highway configuration with open fields, where the strength of the wind is stronger due to the lack of wind shelter by the surroundings. The use of an aerodynamically smooth berm should be considered because the previously discussed negative effects are not predicted there. However, an earth berm requires more space than a conventional barrier, so rows of trees behind noise barriers could be a practical solution in cases with limited space. It also allows "retrofitting" noise walls when wind-related problems have been detected.

In the absence of wind, tall rows of trees will lead to increased downward scattering of sound. Effects are less than ±1 dB(A) for highway noise. The negative effect of downward scattering has to be weighed against the wind direction and wind speed distribution at a given location, and, especially, the orientation of the road, and, consequently, the noise barrier relative to the dominant wind direction is of importance.

Tree canopies can be further "designed," especially by leaving gaps or changing the density of the canopy by species selection. These parameters define where the maximum improvements in sound reduction under windy conditions will be observed. In a successful design, the maximum improvement should coincide with the location where receivers are expected to be located. Similar conclusions, as in the case of a single noise wall,[8] are expected for the double noise barrier case. Gaps and less dense canopies will distribute the positive effects over a larger zone downwind, but the extent of the recovery of shielding that was originally lost by the wind (in absence of trees) will be more limited.

Placing rows of trees behind noise barriers is a common practice, mainly to allow a better integration of vertically erected noise walls in the landscape. This analysis shows that there also are indirect acoustic benefits, namely, preventing and shaping refraction.

2.4.3 Adding vegetated caps on top of conventional noise barriers

In this section, we present results from assessments of the efficiency of vegetated barrier caps (vegetation substrates suitable for growing vegetation) added to an existing 4-m-high rigid barrier. Predictions were conducted for a barrier erected along a four-lane motorway carrying 85% light vehicles (120 km/h) and 15% heavy vehicles (90 km/h). The different barrier cap designs are represented in Figure 2.14.

2.4.3.1 Conclusions

The insertion loss predictions showed that the vegetated caps were efficient only in an area close to the barrier. For such locations, the objective of 5 dB(A) noise reduction at 1.5 m high appears to be feasible by inserting

- a cylinder cap with minimum diameter of 100 cm;
- a T-shape cap (all faces being absorbent) with a minimum width of 90 cm;
- a T-shape cap (with rigid bottom) with a minimum width of 130 cm; or
- a vertical baffles cap (mounted on the barrier with thin elements) with a minimum height of 50 cm.

The objective of 3 dB(A) noise reduction for a 4-m-high receiver is achieved by inserting

- a T-shape cap (all faces being absorbent) with a minimum width of 90 cm;
- a T-shape cap (with rigid bottom) with a minimum width of 130 cm; or
- a vertical baffles cap with a minimum height of 20 cm.

For pedestrians and cyclists close behind the barrier (1 m away), a noise reduction of 5/10 dB(A) is reached by inserting

- a half-cylinder cap (fixed on receiver's side) with minimum diameter of 40 cm/100 cm;

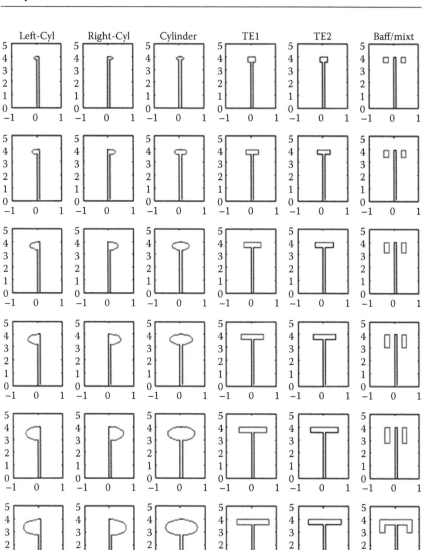

Figure 2.14 Geometries and notations of vegetated caps studied (absorbing substrate in red). Column 1: A ½ cylinder on source side (radius 10–60 cm). Column 2: A ½ cylinder on receiver's side (radius 10–60 cm). Column 3: A ½ cylinder on both sides (diameter 30–130 cm). Columns 4 and 5: T-shaped barriers (50–150 cm wide), the lower side of the top being assumed absorbent or rigid. Column 6: Lateral vertical baffles (20–100 cm), plus a mixed T-shape and baffle case.

- a cylinder cap with minimum diameter of 30 cm/80 cm;
- a T-shape cap (all faces being absorbent) with a minimum width of 50 cm/90 cm; or
- a vertical baffles cap with a minimum height of 20 cm/100 cm.

2.4.3.2 Recommendations

The lower the position of the receiver and the closer it is to the barrier, the more effective is the vegetated cap. So, this mitigation solution should be dedicated to

- pedestrians and cyclists moving close behind the barrier (at most, a few meters away) in order to create a sufficiently noise-shielded path; and
- small recreational areas, gardens, or building entrances situated no more than 20 m behind the barrier.

In the case of half-cylinder caps, we recommend that they should be fixed on the receiver side of the barrier because they offer greater noise reduction in this configuration.

2.4.4 Building low-height earth berms along embanked infrastructure

The aim was to assess the shielding effect of low-height grassy earth berms in open areas (no buildings nearby) whose height is 1 m and cross-sectional area is 1 m². The geometries of the studied berms are given in Figure 2.15 and Figure 2.16.

We considered a four-lane motorway on a 3-m-high embankment in an open area. The modelled traffic scenario was composed of 85% light vehicles at 120 km/h and 15% heavy vehicles at 90 km/h.

2.4.4.1 Conclusions

The extra attenuation due to the addition of a low earth berm along a motorway on a 3-m-high embankment is in the range 3–6 dB(A) for 1.5 m receiver zones and 4–6 dB(A) for the 4-m-high receiver zone. The highest average extra attenuations are obtained for the "square"-shaped B1 and "stairs"-shaped B5 when the lowest one is observed for the "rounded"-shape B4.

Figure 2.15 Geometries of the studied 1-m-high earth berms (B1 to B6).

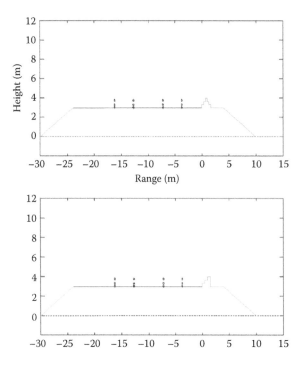

Figure 2.16 General overview for berms B5 and B6 for the motorway configuration.

2.4.4.2 Recommendations

Low-height earth berms proved to be a possible noise-shielding solution for receivers up to 5 m high. We recommend the use of this type of noise abatement solution only when the motorway is embanked (at least a few meters high) or when the altitude of the receiver area is lower by a few meters compared to the infrastructure reference plane.

We also recommend the building of "square" or "stairs" low earth berms (shapes B1, B5, and B6 in Figure 2.15) instead of the more standard "triangle" shapes (B2, B3, and B4).

2.4.5 Building complex-shaped earth berms

The aim was to assess the shielding effect of complex-shaped, 4-m-high earth berms along motorways. Predictions were conducted for a traffic scenario composed of 85% light vehicles at 120 km/h and 15% heavy vehicles at 90 km/h. Note that this shielding performance is expressed as the extra noise attenuation compared to a reference situation: a 4-m-high straight rigid barrier located at the edge of the road or railway infrastructure.

We considered simple nonflat geometries shown in Figure 2.17. The first berm (E1) is a conventional trapezoidal earth berm (red lines). We also carried out optimizations in order to design an optimized noise-reducing device along a motorway by combining an earth berm with a rigid straight or tilted barrier, the overall height of the system being limited to 4 m (an example is shown in Figure 2.18).

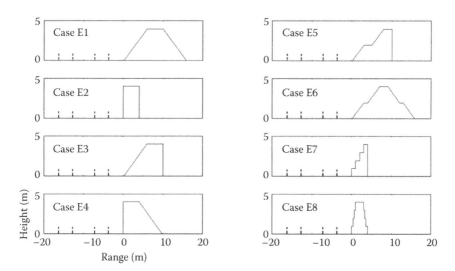

Figure 2.17 Geometry of the eight berms studied for the motorway case (earth with grass is shown in red).

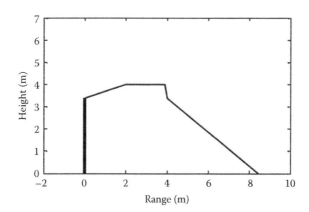

Figure 2.18 An example of the combination of an earth berm (thin line) and straight barrier (thick line).

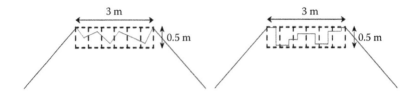

Figure 2.19 An example of triangular and rectangular trough arrangements.

In another work, we studied the acoustical effect of introducing a height profile (in the form of 50-cm-deep troughs) on the top surface of a trapezoidal berm, the troughs being triangular or rectangular as shown in Figure 2.19. Note that the highest points on the profile are below the 4-m-high limit.

We finally studied the effect of fractal roughness (crenel or seesaw profiles) at different levels on one, two, or three faces of a conventional trapezoidal berm.

2.4.5.1 Conclusions

In the case of simple nonflat geometries (shown in Figure 2.17), the extra attenuation is less than in the railway case. Shapes E2 and E4 show the highest values, up to 6 dB(A). On the other hand, shapes E1, E3, and E6 are predicted to be less efficient than the reference straight barrier by up to 3 dB(A) in the case of E1 (conventional trapezoidal berm) and 5 dB(A) for E6 (double-slope trapezoidal barrier).

One may note that for the "pedestrian" receiver, the conventional trapezoidal shape E1 and E6 are predicted to have the worst performance with an acoustical attenuation of 3–5 dB(A) less than a reference straight barrier, when best berms are E7 and E8 (stairs).

For noise systems combining an earth berm with a rigid barrier, the highest extra attenuation obtained is 6 dB(A) for a straight barrier and 3 dB(A) for a tilted one (compared to a reference 4-m-high straight rigid barrier).

Introducing 50-cm-deep nonflatness on top of a conventional trapezoidal berm may lead to an improvement up to 4 dB(A) for rectangular troughs and 2 dB(A) for triangular ones.

Concerning the addition of fractal roughness (studied for 1.5-m-high receiver zone 3), the most important berm face to treat is the top. Results show that the extra noise attenuation (compared to a conventional smooth berm) obtained when adding roughness on the top face may only reach 3, 5, and 7 dB(A) (crenel profiles and roughness level 1, 2, and 3, respectively). Roughening the two other faces just brings 1–2 dB(A) more. Seesaw profiles are less efficient.

2.4.5.2 Recommendations

We recommend that earth berms along motorways be designed to have

- an E4 shape (the "source" face of the berm is vertical, or almost vertical. This can be achieved by combining a barrier or gabions with an earth berm); and
- E5 or E6 shapes (nonflatness is added to obtain a "stairs" shape).

Because the extra attenuation at the "pedestrian" receiver may be increased by 5–6 dB(A), this action is recommended to reduce noise exposure for those walking or cycling just behind the berm.

If treatment of the berm top is possible, we recommend creating large and deep troughs that are predicted to yield higher performance than shallow ones. In creating this type of roughness, we recommend the use of crenel shapes instead of seesaw profiles.

2.4.6 Building low-height vegetated barriers at the edges of bridges

We assessed the extra noise attenuation due to a 1-m-high absorbing barrier (consisting of a 40-cm-wide vegetation substrate surrounding a thin rigid core), compared with a nonabsorbing barrier. Predictions were conducted for one or several barriers built along one or both edges of a 6-m-high bridge carrying a four-lane motorway as shown in Figure 2.20(a)(b). The traffic composition was assumed to be of 85% light vehicles at 120 km/h and 15% heavy vehicles at 90 km/h.

We focus here on a pedestrian or a cyclist moving below the bridge, both receivers being at a height of 1.5 m (dotted lines).

2.4.6.1 Conclusions

The 40-cm-wide, 1-m-high barrier made from vegetation substrate over a rigid core is very efficient for improving the noise environment below the bridge where pedestrians and cyclists are moving. Even when no inner rigid core is used inside the barrier, a high noise-reducing effect is predicted for both configurations. The extra noise-reducing effect (in comparison with a situation without barrier) is in the range 6–8 dB(A).

2.4.6.2 Recommendations

The shielding effect of adding a low absorbing barrier at the edge of bridges can be quite large. We recommend the use of such low barriers to improve the noise environment below bridges, notably where pedestrian and cycle paths have been designed.

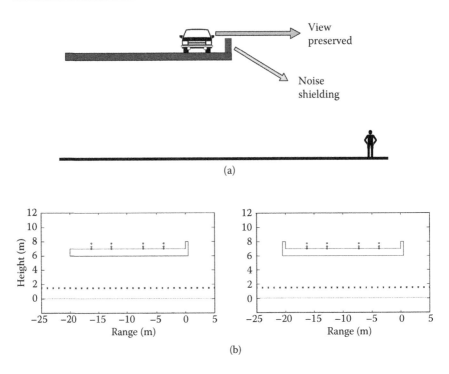

Figure 2.20 (a) An illustration of the effect of a bridge barrier, and (b) geometries used for calculations with barrier on one (bottom left) or both sides (bottom right).

One may note that the addition of a 1-m-high absorbing barrier at the edge of bridges does not prevent car drivers from seeing the landscape beyond the bridge. Moreover, the light weight of the vegetation substrate makes the product suitable for bridges. By creating noise-shielding areas below bridges, we aim at promoting environmentally friendly transportation modes (cycling, walking) as well as urban parks and recreational areas.

2.5 RAILWAYS

2.5.1 Building low-height earth berms along embanked infrastructures

The aim was to assess the shielding effect of low-height grassy earth berms in open areas (no building nearby) whose height is 1 m and cross-sectional area is 1 m². The geometries of the studied berms are given in Figure 2.21.

Figure 2.21 Geometries of the studied 1-m-high earth berms (B1 to B6).

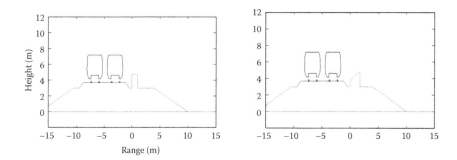

Figure 2.22 General overview for berms B1 and B2 for the TGV railway configuration.

We consider a two-track, high-speed train (TGV high-speed trains moving at 300 km/h) on a 3-m-high embankment. Note that the berms are 1 m high above the ballast plane (considered to be 50 cm thick), so that their total height is 1.5 m, as shown in Figure 2.22.

2.5.1.1 Conclusions

The extra attenuation due to the addition of a low earth berm along a two-track TGV railway on a 3-m-high embankment is in the range 4–8 dB(A) for 1.5-m receiver zones and 5–9 dB(A) for the 4-m-high receiver zone. The highest average extra attenuations are obtained for the "square" and "stairs" shapes B2, B5, and B6; the lowest ones for the "slope" shape B3.

2.5.1.2 Recommendations

Low-height earth berms proved to be a potential noise-shielding solution for receiver positions up to 5 m high. We recommend use of this type of noise abatement solution only when the TGV railways are embanked (at least a few meters high) or when the altitude of the receiver area is lower by a few meters compared to the infrastructure reference plane.

We recommend building "square" or "stair-shaped" low earth berms (shapes B1, B5, and B6 in Figure 2.21) instead of the more conventional triangular shapes (B2, B3, and B4).

2.5.2 Building complex-shaped earth berms

The aim was to assess the shielding effect of complex shaped, 4-m-high earth berms along railways (TGV high-speed train moving at 300 km/h or freight moving at 80 km/h). Note that this shielding performance was expressed as the extra noise attenuation compared to a reference situation: a 4-m-high, straight rigid barrier located at the edge of the railway infrastructure.

We considered simple nonflat geometries shown in Figure 2.23. The first berm (E1) is a conventional trapezoidal earth berm.

2.5.2.1 Conclusions

Along railways where freight trains are moving, the extra noise attenuation (compared to a reference 4-m-high straight rigid barrier) is in the range 3–7 dB(A) for the 1.5-m-high receiver zones, and 4–6 dB(A) for the 4-m-high receiver zones. The shapes predicted to yield the highest performance are E2, E3, and E4 (see illustration in Figure 2.23).

In the case of a TGV moving on the railway, the extra noise attenuation was less; in the range 2–5 dB(A), the shapes predicted to yield the highest performance were E2, E4, and E5.

One may note that for the "pedestrian" receiver, the conventional trapezoidal shape E1 is predicted to have the worst performance, with an acoustical attenuation 2 dB(A) less than a reference straight barrier.

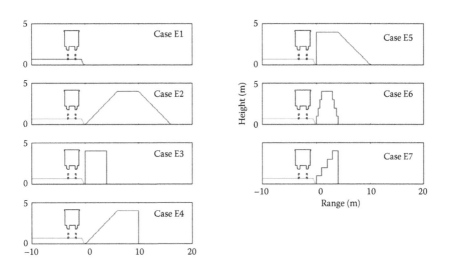

Figure 2.23 Geometry of the no berm and the seven berms studied for the trains case (earth with grass in red, ballast in green).

2.5.2.2 Recommendations

The 4-m-high berms were shown to be more useful next to railways than motorways. We recommend that earth berms are designed to have an E4 shape (the "source" face of the berm is vertical, or almost vertical). This can be achieved by combining a barrier with an earth berm.

Because the extra attenuation at the "pedestrian" receiver may be increased by 5–6 dB(A), this action is recommended to reduce noise exposure in areas where there are people walking or cycling just behind the berm.

2.6 SUMMARY OF CONCLUSIONS

This research on innovative noise barriers shows that families of innovative barriers using natural materials are effective, promising solutions to abate ground transportation noise for receivers up to 5 m high. These novel solutions depend on both transportation and environment types.

From this work one can draw the following conclusions:

- The initial objectives (see Table 2.1) can be achieved in most cases:
  - *In urban streets*: By using low vegetated barriers and low gabions (efficient for low receivers and when they are not too close to the barrier)
  - *Along tramways*: By using low vegetated barriers, low berms, and sonic crystal-assisted barriers
  - *Along motorways on flat terrain*: By using vegetated caps (for close receivers) and complex shape berms
  - *Along motorways on an embankment*: By using vegetated caps and low berms
  - *Along railways on flat terrain*: By using complex shape berms
  - *Along railways on an embankment*: By using low berms.
- Vegetation substrate is predicted to be very suitable for use on low barriers and vegetated caps (planted vegetation has no significant effect on the substrate's sound-absorbing properties).
- A row of trees (with an adapted canopy design) behind a conventional noise barrier is an option to maintain the acoustic shielding provided by noise barriers in downwind conditions. It is especially interesting in a highway configuration in open fields where winds can be strong.
- The noise abatement achieved by adding an extra interlane (streets) or intertrack (tramways) absorbing low barrier is considerable. This is a highly promising solution to create noise-shielded areas and walkways in cities.

- The use of classic, smooth, trapezoidal shapes for earth berms along rural transportation corridors does not appear to be an efficient noise abatement solution; other shapes could be considered.
- Low vegetated barriers at the edge of bridges seem to be promising, easy-to-implement solutions to improve the soundscape for walking and cycling paths below them.

REFERENCES

1. Defrance, J., P. Jean, F. Koussa, T. Leissing, K. Attenborough, S. Taherzadeh, D. Botteldooren, T. Van Renterghem, K. Horoshenkov, A. Khan, H. Benkreira, J. Forssén, J. Kang, A. Mandon, and R. Rohr. 2011. *State of the art of experience and* models (FP7 HOSANNA deliverable 2.1). Online at: http://www.greener-cities.eu/

2. Defrance, J., K. Attenborough, K. Horoshenkov, F. Koussa, P. Jean, and T. Van Renterghem. 2011. *Choice and adaptation of models* (FP7 HOSANNA deliverable 2.2). Online at http://www.greener-cities.eu/

3. Defrance, J., P. Jean, F. Koussa, K. Horoshenkov, A. Khan, H. Benkreira, T. Van Renterghem, J. Kang, Y. Smyrnova, and J. Forssén. 2013. *Application to innovations* (FP7 HOSANNA deliverable 2.3). Online at: http://www.greener-cities.eu/

4. Defrance, J., P. Jean, F. Koussa, K. Horoshenkov, A. Khan, H. Benkreira, T. Van Renterghem, J. Kang, and Y. Smyrnova. 2013. *Technical report with recommendations for innovative barriers* (FP7 HOSANNA Deliverable 2.4). Online at: http://www.greener-cities.eu/

5. Nota, R., R. Barelds, and D. Van Maercke. 2005. *Harmonoise WP 3 engineering method for road traffic and railway noise after validation and fine tuning.* Deliverable of WP3 of the HARMONOISE project. Document ID HAR32TR-040922-DGMR20.

6. Horoshenkov, K., A. Khan, H. Benkreira, A. Mandon, and R. Rohr. 2014. Acoustic performance of vegetation and soil substratum in an urban context. In *Environmental methods for transport noise reduction*, eds. M. E. Nilsson, R. Klæboe, and J. Bengtsson (pp. 49–80). Oxford, UK: Spon Press.

7. Van Renterghem, T., K. Attenborough, and P. Jean. 2014. Designing vegetation and tree belts along roads. In *Environmental methods for transport noise reduction*, eds. M. E. Nilsson, R. Klæboe, and J. Bengtsson (pp. 93–120). Oxford, UK: Spon Press.

8. Van Renterghem, T., P. Jean, and J. Defrance. 2012. *Improving microclimatology by trees, shrubs, and bushes* (FP7 HOSANNA Deliverable 3.3). Online at: http://www.greener-cities.eu/

Chapter 3

Acoustic performance of vegetation and soil substratum in an urban context

Kirill Horoshenkov, Amir Khan,
Haidj Benkreira, Agnes Mandon, and René Rohr

Abstract. This chapter presents the measured and predicted data for the acoustic absorption coefficient of soil with and without living plants. These data show that the ability of soil to absorb sound is controlled by the soil porosity and the amount of moisture in the soil. It is shown that these effects can be predicted using an equivalent fluid model and transfer matrix approach. The same modelling approach can be used to predict the absorption of a plant in the absence and presence of soil. The ability of a living plant to absorb sound is controlled by the plant height, leaf area density (LAD), and dominant angle of leaf orientation. Generally, plants with a higher LAD and angle of leaf orientation are able to absorb a higher proportion of the incident sound energy and across a broader frequency range. It is shown that living plants can significantly enhance the absorption of clay-based soils with relatively low proportion of open, interconnected pores across the entire frequency range of interest. In the case of light density soil, the presence of a living plant enhances the ability of soil to absorb sound in the low and high frequency ranges. The diffused absorption coefficient of these plant–soil systems can be designed to be similar or higher to that expected from traditional manmade porous absorbers through a careful selection of the type of living plant.

3.1 INTRODUCTION

Foliage and hedges have many important functions for our environment, especially in densely populated urban environments. They produce oxygen, purify the air, and serve as visual aesthetic elements, windbreaks, and habitats for birds and insects. Plants grown on a building façade (Figure 3.1) contribute to balanced microclimate by their ability to store humidity. However, plants are rarely used for noise control in urban development because it has long been believed that they possess no useful acoustical properties. The aim of the research presented in this chapter is to investigate the potential of using plants in noise control elements and to determine

Figure 3.1 Photograph of a living façade in Lyon, France.

those characteristics of plants that make them particularly good in terms of their sound absorption and transmission properties. The reported investigations have been performed under controlled laboratory conditions as well as outdoors. In addition, theoretical and empirical models have been adopted and applied to explain the observed attenuation and absorption coefficients and to predict the acoustic properties of plants when used as a part of urban and landscape architecture.

The first section of this chapter examines the experimental setup and measurement procedures; how the plant soil systems are studied in an impedance tube. The theory behind the impedance measurements is explained. The next section compares two different soil types and their performance when saturated. The section that follows studies the response of a model for soils with varying saturation levels. The study of low-growing plants forms the final section in which the plant soil systems are modelled using the transfer matrix approach.

3.2 EXPERIMENTAL SETUP AND MEASUREMENT PROCEDURES

To design an effective green noise-control element, it is necessary to study the interactions between the sound field in the plant and the porous soil that supports the growth of this plant. It is natural to assume that these interactions are controlled by the leaf area, leaf area density, soil porosity, pore size, and moisture distribution in the soil.

A simple way to determine the ability of plants to absorb sound is to measure its plane wave, the normal incidence absorption coefficient. This can be achieved by placing a single plant specimen with or without soil in a sufficiently large impedance tube (e.g., ISO[1]). This experimental arrangement is illustrated in Figure 3.2, which shows how the soil only, plant only, and plant on top of soil can be introduced in the impedance tube.

Because of the random geometry of a living plant specimen, it is impossible to determine the exact position of the reference plane from which the distance from the top of the plant (x_1) to the farther microphone location can be accurately measured (see Figure 3.2 (b)(c)). Therefore, it is possible to measure directly only the absolute value of the acoustic reflection coefficient, r, which is given as:

$$r = |r| e^{i\varphi} = \frac{H_{12} - H_1}{H_R - H_{12}} e^{2ik_0 x_1}, \tag{3.1}$$

where H_{12} is the measured transfer function between the two microphones, H_I is the transfer function for the incidence wave between the two microphone positions, H_R is the transfer function for reflected wave,

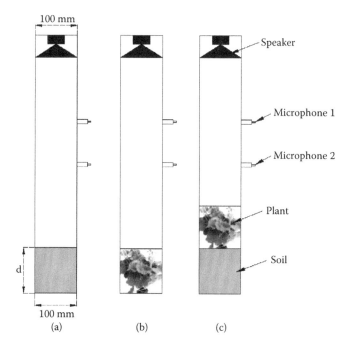

Figure 3.2 Schematic illustration of the impedance tube with (a) soil only, (b) plant only, and (c) plant placed on top of soil.

and K_0 is the wavenumber. Because of the ambiguity in the position of the reference place, it was impossible to determine the term $i\phi$ in Equation (3.1), which is needed to calculate the complex acoustic surface impedance of the plant specimen.[1] Therefore, the following sections will only deal with the acoustic absorption coefficient, which is calculated according to the following expression:

$$\alpha = 1 - |r|^2, \tag{3.2}$$

where α is the acoustic absorption coefficient.

3.3 EFFECT OF MOISTURE ON SOIL ABSORPTION

Plants live in soil from which they draw nutrients and water to support their growth. It has been shown in previous works (e.g., Horoshenkov and Mohamed[2]) that the amount of soil moisture can significantly affect its acoustical properties. This section aims to illustrate the influence of moisture on acoustic absorption in two types of soils: (1) high-density clay-based soil used in ordinary flower beds, and (2) light-weight substratum soil developed by Canevaflor (France) for use in vertical living walls. The acoustic absorption coefficient for these soils is examined here through laboratory studies where water content was systematically varied.

It can be argued that the proportion of open, interconnected pores (porosity), pore size distribution, and resulting air flow resistivity are the key characteristics that control the acoustic properties of the plant–soil system. Here we examine the acoustical and related nonacoustical properties of two types of soil that can be used as a plant habitat. These soils are a light-density soil substratum supplied by Canevaflor and a high-density clay-based soil that is used for garden planting across the United Kingdom. Figure 3.3 are photos of these soils. The porosity, ϕ, of a typical clay-based soil falls between $0.26 < \phi < 0.48$, depending on the packing. Low-density soil substratum, such as that supplied by Canevaflor, has a much higher porosity of $0.70 < \phi < 0.85$. This material consists of perlite, an amorphous volcanic glass that has a relatively high water content, coconut fibres, and hydrophilic polymer gel. Compared to clay-based soil, little or no compaction occurs in low-density soil and this substratum is able to retain at least three times its own weight of water due to the presence of hydrophilic polymer gel. The texture and the amount of conserved moisture determine the pore size distribution, which controls the acoustical properties of soil. In these experiments, the soil samples were prepared for uniformity throughout the depth. This was ensured by taking five cores from the soil and mixing them thoroughly to represent the average conditions within an area. The soil was then carefully weighed to ensure that the soil density was

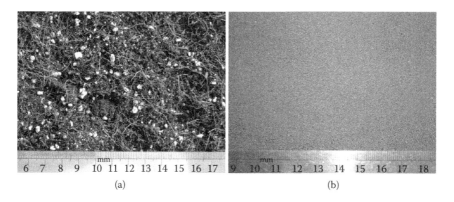

6 7 8 9 10 11 12 13 14 15 16 17   9 10 11 12 13 14 15 16 17 18
        (a)   (b)

Figure 3.3 (a) Substratum soil and (b) clay-based soil.

consistent from sample to sample and placed inside the sample holder for testing. The density of substratum soil is typically 250 kg/m³, whereas the density of clay-based soil is in excess of 1,200 kg/m³.

In order to illustrate the effect of moisture on the absorption coefficient of soil, a controlled amount of water can be sprayed on the surface of the soil sample, which is loaded in a specially manufactured sample holder. This sample holder can then be left for a period of time, which is sufficient for the added water to become fully absorbed by the soil sample before the acoustic absorption coefficient of it can be measured with the help of an impedance tube, shown in Figure 3.2(a).

Generally, there are three major effects on the acoustic absorption that are associated with the moisture present in soil. These relate to (1) reduction in soil porosity, (2) increase in the soil flow resistivity, and (3) soil aggregation stimulated by the coating of clay particles coated in the meniscus of water. These changes tend to occur at the top few centimetres of a layer of soil when water is applied externally. A reduction in the pore size is observed as a result, and it is reflected in the acoustic absorption coefficient of the soil. This effect is illustrated in Figure 3.4, which shows an apparent reduction in the acoustic absorption coefficient when $V_W = 100$ cm³ of water was added to a $V_T = 1.57$ litre (200-mm-thick and 100-mm-diameter) soil specimen. Because of the porosity differences in clay soil and substratum soil, the addition of a relatively small amount of water has markedly different effects on the acoustic absorption of these two types of soils. The level of reduction in the absorption coefficient can be explained by the change in the degree of saturation (S_r) to which these two samples were subjected. The degree of saturation is defined by the following expression:

$$S_r = V_W / V_v,$$

(3.3)

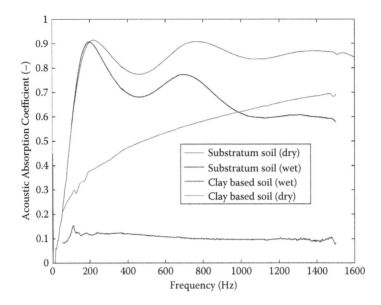

Figure 3.4 The effect of the degree of saturation on the absorption coefficient of substratum and clay-based soils.

where the volume of voids is

$$V_v = V\phi, \tag{3.4}$$

V being the total volume of the material specimen.

It is clear that substratum soil is able to retain a great deal more moisture than clay-based soil because of its higher porosity (two times higher than that of clay). This is explained by the presence of the porosity term in the denominator of Equation (3.3). Therefore, the addition of 100 cm^3 of water to the substratum soil results in a relatively small, 9%, increase in the degree of saturation. In the case of clay-based soil, the addition of 100 cm^3 of water results in a relatively large, 17%, increase in water saturation.

The change in the absorption coefficients shown in Figure 3.4 at different frequencies are different for the two soil specimens. In the case of clay-based soil, there is an almost seven-fold overall reduction in the observed absorption coefficient across the entire frequency range. In the case of low-density soil substratum, this change is relatively small (maximum 1.5-fold) and it is confined to the higher frequency range only. This behaviour enables us to draw the following conclusions: (1) in the case of clay-based soil with smaller pores, the effect of added water is more profound and the moisture distribution more uniform across the entire specimen; and (2) the effect of added water in the case of light-density soil substratum is more complex and seems to be confined to the top layer of soil through which the water was administered.

3.4 MODELLING OF THE ACOUSTICAL PROPERTIES OF SOILS

The results presented in the previous section provide a basis for the selection of suitable modelling methods to predict the fundamental acoustical properties of soils, such as the characteristic impedance and complex wavenumber. Dry soil or soil with a low degree of saturation is characterised by a relatively uniform pore size distribution. This enables us to use a three-parameter model proposed by Miki,[3] which accounts for a uniform pore size and a limited porosity. This model also may work well in the case of clay-based soil with a relatively high degree of saturation, S_r, provided that the moisture distribution is relatively uniform. Clay-based soils consist of pores that are relatively small in size (around 40–60 μm diameter). Therefore, there are very strong capillary forces in this type of porous media that help to achieve a relatively uniform distribution of moisture. The performance of the Miki model (see Section 3.6 for model details) is illustrated in Figure 3.5, where the measured and predicted values of

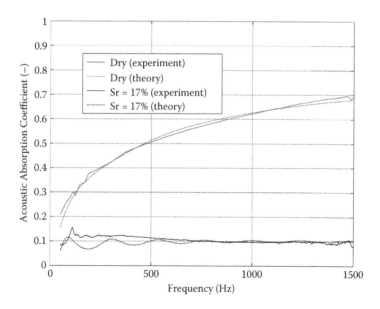

Figure 3.5 Measured and predicted absorption coefficients of clay-based soil in different saturation states.

the absorption coefficient are presented for a 200-mm-thick sample of clay-based soil in different degrees of water saturation. In this model, the following values of the nonacoustical parameters were used: (1) flow resistivity σ = 47 kPa s m^{-2}, ϕ = 0.45, and tortuosity α_∞ = 1.00 for dry soil; and (2) σ = 110 kPa s m^{-2}, ϕ = 0.1, and tortuosity α_∞ = 11.15 for wet soil. There is a good agreement between the measured and predicted values of the absorption coefficient across the entire frequency range.

On the other hand, porous media, such as low-density soil substrate, consist of pores of a relatively large diameter (200 μm and above). This soil also contains a proportion of meso-pores (>1 mm scale), which are created around the particles of perlite and polymer gel that are relatively large. If the top surface of this material is sprayed with a relatively small amount of water (e.g., S_r < 10%), then one can expect that this moisture will only penetrate into the first few millimetres of the sample just to be absorbed by the polymer gel particles confined to the top. This phenomenon results in a material with pore size stratification. The acoustical properties of this multilayered material can no longer be modelled using a uniform pore size distribution assumption and pore size stratification information is required.

A relatively straightforward method to account for this stratification is to use a two-layer model. In this way, the acoustical properties of the top, water-saturated layer and the bottom, relatively dry layer of soil can be modelled using two sets of nonacoustical parameters. The surface impedance of a stack of two layers with difference characteristic acoustic impedances and propagation constants is predicted by the following expression:

$$Z(\omega) = \tilde{Z}_t \frac{\tilde{Z}_b + \tilde{Z}_t \tanh(-i\tilde{k}_t d_t)}{\tilde{Z}_t + \tilde{Z}_b \tanh(-i\tilde{k}_t d_t)}, \tag{3.5}$$

where $\tilde{Z}_t, \tilde{K}_t$ are the characteristic impedance and wavenumber in the top layer, respectively, and d_t is the thickness of the top layer. In the above expression, the input impedance of the bottom layer is $Z_b = \tilde{Z}_b \coth(-i\tilde{K}_b d_b)$, where $\tilde{Z}_b, \tilde{K}_b$ are the characteristic impedance and wavenumber in the bottom layer, respectively. Similarly, the variable d_b corresponds to the thickness of the bottom layer. An optimisation procedure similar to that detailed by Horoshenkov and Swift[4] was used to derive the values of d_b, d_t, and four other nonacoustical parameters that are required to predict the acoustical absorption coefficient of soil substrate at S_r = 9%. These parameters were the flow resistivity of the top layer, flow resistivity of the bottom layer, porosity of the bottom layer, and its tortuosity. The porosity and pore size distribution of the top layer was assumed to be constant and equal to the mean values obtained for the case of the dry material specimen. The model proposed by Horoshenkov et al.[5] was employed together with Equation (3.5) to predict the values of $\tilde{Z}$ and $\tilde{K}$. In the case of dry soil substrate, we adopted a standard single-layer model[5] and assumed that the pores are triangular in shape and their size distributed log-normally with the log-normal measure of the standard deviation, $\bar{s}$.

Figure 3.6 presents the measured and predicted absorption coefficient spectra for dry soil substrate and for soil substrate with 100 cm^3 water applied to its surface ($S_r = 9\%$). The values of nonacoustical parameters for dry soil substratum were: (1) flow resistivity $\sigma = 6.87$ kPa s m^{-2}, porosity $\phi = 0.72$, tortuosity $\alpha_\infty = 1.89$, and standard deviation in pore size $\bar{s} = 0.3$ for the dry, uniform $d = 200$-mm-thick layer of soil substratum; and (2) $\sigma_t = 24.0$ kPa s m^{-2}, $\sigma_b = 2.84$ kPa s m^{-2}, $\phi_t = 0.70$, $\phi_b = 0.72$, $\alpha_\infty^{(t)} = 1.00$, $\alpha_\infty^{(b)} = 1.20$, $\bar{S}_t = \bar{S}_b = 1.10$, and $d_t = 0.0075$ m, $d_b = 0.1925$ m for the stratified layer of wet soil substratum. Here, the indices t and b stand for the top and bottom layers, respectively. Here, the nonacoustic parameters were recovered using an optimisation technique because it was impossible to measure them directly.

The values obtained from the optimisation stage produced parameter values that made good physical sense. They signify that the flow resistivity of the top layer of soil substrate is considerably higher than that of the bottom layer. This is consistent with the moisture near the top. The optimisation results can be explained by the porosity of the bottom layer not being affected by moisture. However, the pore size distribution in the top and bottom layers increase because the added water has increased the size of

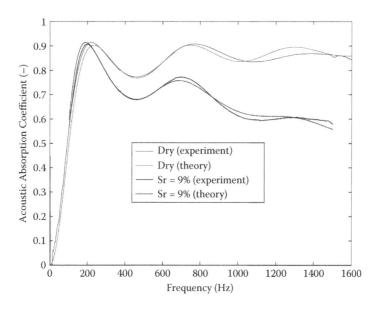

Figure 3.6 Measured and predicted absorption coefficients of low-density soil substratum in different saturation states.

the particles of polymer gel. The flow resistivity of the top layer is 10-fold greater than that of the bottom. The estimated thickness of the top layer is 7.5 mm, which is comparable with the size of polymer gel particles in their expanded state. In the case of the dry soil substrate, it is sufficient to assume that pore size distribution is uniform along the material specimen so that the standard four-parameter model[5] can be adopted.

3.5 LOW GROWING PLANTS

To understand better the relationship between the absorption coefficient of a plant and some basic leaf characteristics, we can examine five specimens of each of five different plant species that have a representative range of leaf morphologies, areas, and orientations. These five plant species, purchased from a garden centre, include *Geranium zonale*, *Hedera helix* (Green Ivy), *Pieris japonica*, *Summer Primula vulgaris* (Summer Primrose), and *Winter Primula vulgaris* (Winter Primrose) (Figure 3.7). The photos in Figure 3.8 show leaves that were collected from each of these five types of plants. This figure illustrates that the morphology of the plants studied is sufficiently different from plant to plant in terms of leaf shape and dimensions. It is of interest to measure and analyse the following plant leaf characteristics: mean thickness of a single leaf (h_f), mean weight of a single leaf (w_f), mean area of a single leaf (a_f), number

(a)   (b)   (c)

(d)   (e)

Figure 3.7 Five types of plant species used in the acoustic experiments: (a) *Geranium zonale*, (b) *Hedera helix*, (c) *Pieris japonica*, (d) *Summer Primula vulgaris*, and (e) *Winter Primula vulgaris*.

Figure 3.8 Photographs of leaves from the five types of plants used in this work:
(a) *Geranium zonale*; (b) *Hedera helix*; (c) *Pieris japonica*; (d) *Summer Primula vulgaris*; (e) *Winter Primula vulgaris*.

of leaves on a plant (n_f), equivalent height of the plant (h_p), equivalent volume occupied by the plant (V_p), and dominant angle of leaf orientation (θ_f). These characteristics can be determined according to the procedure described by Horoshenkov et al.[6]

Table 3.1 provides a summary of the key characteristics of the plant foliage that can be used to estimate the total area of leaves on the plant (S_f), the surface density of a leaf (ρ_f), the total weight of leaves on the plant (W_p), and the leaf area density (A_v). Table 3.2 presents a summary of the plant characteristics derived from these data.

3.6 MODELLING OF THE ACOUSTICAL PROPERTIES OF PLANTS

In order to understand the acoustical properties of a plant, it is convenient to represent it in the form of an equivalent fluid, because it is also common to predict the acoustical properties of soil with a model that treats soil as an equivalent fluid medium.[3] In this case, the acoustical absorption of a plant–soil system can be determined using a transfer matrix approach (see Equation (3.5)) provided the characteristic impedance, $\tilde{Z}$, and wavenumber, $\tilde{K}$, for each of

Table 3.1 The basic morphological characteristics of plant leaf specimens

Plant species	Average thickness of single leaf, (mm) h_f	Average weight of single leaf, (g) w_f	Average area of single leaf, (m²) a_f	Number of leaves on plant, n_f	Estimated height of plant, (m) h_p	Equivalent volume occupied by plant, (m³) V_p	Leaf area density, (m⁻¹) A_v	Dominant angle of leaf orientation, (Degrees) θ_f
Geranium zonale – 1	0.55	2.57	0.0044	15	0.20	0.0016	42	85(±5)
Geranium zonale – 2	0.68	2.40	0.0043	14	0.17	0.0013	45	
Geranium zonale – 3	0.64	2.56	0.0041	13	0.17	0.0013	40	
Geranium zonale – 4	0.65	2.56	0.0044	15	0.20	0.0016	42	
Geranium zonale – 5	0.63	1.53	0.0027	16	0.18	0.0014	30	
Hedera helix – 1	0.36	0.13	0.00062	83	0.27	0.0021	24	36(±10)
Hedera helix – 2	0.31	0.13	0.00064	92	0.17	0.0013	44	
Hedera helix – 3	0.36	0.08	0.00066	83	0.17	0.0013	41	
Hedera helix – 4	0.38	0.13	0.00072	110	0.17	0.0013	59	
Hedera helix – 5	0.31	0.06	0.00039	59	0.17	0.0013	17	
Pieris japonica – 1	0.52	0.17	0.00038	362	0.28	0.0022	63	30(±20)
Pieris japonica – 2	0.53	0.24	0.00061	366	0.33	0.0026	86	
Pieris japonica – 3	0.53	0.15	0.00032	228	0.27	0.0021	34	
Pieris japonica – 4	0.58	0.24	0.00045	526	0.33	0.0026	91	
Pieris japonica – 5	0.51	0.24	0.00062	320	0.29	0.0023	87	
SPrimula vulgaris – 1	0.35	3.20	0.0066	18	0.23	0.0018	66	14(±15)
SPrimula vulgaris – 2	0.31	2.55	0.0056	21	0.28	0.0022	54	
SPrimula vulgaris – 3	0.30	3.60	0.0063	32	0.24	0.0019	107	
SPrimula vulgaris – 4	0.27	3.82	0.0067	19	0.28	0.0022	58	
SPrimula vulgaris – 5	0.26	2.12	0.0043	24	0.25	0.002	52	
WPrimula vulgaris – 1	0.50	2.19	0.0051	22	0.1	0.0008	143	72(±15)
WPrimula vulgaris – 2	0.51	1.60	0.0054	19	0.095	0.0007	138	
WPrimula vulgaris – 3	0.59	2.03	0.0053	22	0.11	0.0009	136	
WPrimula vulgaris – 4	0.57	1.93	0.005	16	0.1	0.0008	102	
WPrimula vulgaris – 5	0.53	2.83	0.0066	17	0.09	0.0007	159	

Table 3.2 The parameters of plant leaf specimens derived from the data provided
in Table 3.1

Plant species	Total area of leaves on plant, S_f (m^2)	Surface density of single leaf, ρ_f (kg/m^2)	Total weight of leaves/ stems, W_p (kg)	Leaf area density, A_v (m^{-1})
Geranium zonale – 1	0.066	0.58	0.045	42
Geranium zonale – 2	0.060	0.56	0.045	45
Geranium zonale – 3	0.053	0.62	0.039	40
Geranium zonale – 4	0.066	0.54	0.047	42
Geranium zonale – 5	0.043	0.57	0.034	30
Hedera helix – 1	0.050	0.21	0.017	24
Hedera helix – 2	0.059	0.20	0.023	44
Hedera helix – 3	0.055	0.12	0.021	41
Hedera helix – 4	0.079	0.18	0.021	59
Hedera helix – 5	0.023	0.15	0.013	17
Pieris japonica – 1	0.138	0.45	0.081	63
Pieris japonica – 2	0.223	0.39	0.115	86
Pieris japonica – 3	0.073	0.47	0.053	34
Pieris japonica – 4	0.237	0.53	0.112	91
Pieris japonica – 5	0.198	0.39	0.084	87
SPrimula vulgaris – 1	0.119	0.48	0.074	66
SPrimula vulgaris – 2	0.118	0.46	0.074	54
SPrimula vulgaris – 3	0.202	0.57	0.110	107
SPrimula vulgaris – 4	0.127	0.57	0.063	58
SPrimula vulgaris – 5	0.103	0.49	0.048	52
WPrimula vulgaris – 1	0.112	0.43	0.059	143
WPrimula vulgaris – 2	0.103	0.30	0.057	138
WPrimula vulgaris – 3	0.117	0.38	0.076	136
WPrimula vulgaris – 4	0.080	0.39	0.055	102
WPrimula vulgaris – 5	0.112	0.43	0.054	159

the two equivalent layers in this system are known. The use of this model
enables consideration of the viscous losses and thermal dissipation of the
sound wave as it propagates through the plant foliage and through the layer
of porous soil. It is believed that these mechanisms can explain the acoustic
absorption coefficient spectra observed in the impedance tube experiments
with the plants.

As shown in the previous sections, the empirical model proposed by
Miki[3] is able to provide satisfactory predictions of the fundamental acoustic
properties of soils. In this section, we will use the same model to predict
the acoustical properties of low-growing plants. It can be shown that this

model relies on the porosity and tortuosity that can be measured directly and linked to the morphology of the plant.

Taking these two parameters into account improves the accuracy of the predicted acoustic absorption coefficient of a porous medium at the medium and high frequencies. It is easy to demonstrate that the same close fit cannot be achieved with a single parameter model based only on the equivalent flow resistivity.

The porosity of a plant can be estimated from:

$$\phi = V_f / V_p, \tag{3.6}$$

where $V_f = S_f h_f$ is the volume of the plant foliage and V_p the total equivalent volume occupied by the plant.

The porosity of soil can be estimated using the procedure detailed by Leclaire et al.[7] In the case of soil, it is feasible to measure the flow resistivity and tortuosity directly and accurately using the standard procedure defined by ISO 9053.[8] However, the measured value of the flow resistivity of loose soil *in situ* depends strongly on the degree of soil compaction, which is difficult to match exactly with that achieved when the same soil specimen is placed in the impedance tube. The variability between what is measured in the lab and in the field can be rather high.

Measurements of the flow resistivity of a plant in a laboratory setting are problematic because it is impossible to estimate accurately the thickness of the plant inserted in the flow resistivity tube, and the pressure drop across the plant specimen is relatively small to be measured accurately for the flow velocity range recommended in the ISO 9053 standard.[8] The tortuosity of a plant also is impossible to measure precisely because existing measurement methods require the exact thickness of the material specimen.

In order to overcome these problems, the measured acoustic absorption coefficient for each individual plant specimen can be used as input data in a standard optimisation algorithm. For this purpose, it is possible to adopt the following objective function to minimise the following difference:

$$F(x) = \sum_{n=1}^{N} \left| \alpha^{\exp}(f_n) - \alpha^{th}(f_n, \mathbf{x}) \right| \rightarrow \min, \tag{3.7}$$

where $\alpha^{exp}(f_n)$ is the measured absorption coefficient spectrum, $\alpha^{th}(f_n, X, \phi)$ the predicted absorption coefficient spectrum, f_n the frequency of sound, and $\mathbf{x} = \{\sigma, \alpha_\infty, d_e\}$ the design variable vector in which d_e is the equivalent height of the plant.

In the case of the soil, the design variable vector was $\mathbf{x} = \{\sigma, \alpha_\infty\}$ because the soil layer thickness was accurately determined. The optimisation procedure used to minimise the objective function, Equation (3.7), can be based on the Nelder–Mead direct search optimisation method.[9]

In this work, we use the Miki model.[3] This model suggests the following expressions for the characteristic impedance:

$$Z_b(f) = \frac{\sqrt{\alpha_\infty}}{\phi}\left\{1 + 0.070\left(\frac{f}{\sigma}\right)^{-0.632} + 0.107i\left(\frac{f}{\sigma}\right)^{-0.632}\right\} \tag{3.8}$$

and the complex wavenumber:

$$K_b(f) = \frac{2\pi f\sqrt{\alpha_\infty}}{c_0}\left\{1 + 0.109\left(\frac{f}{\sigma}\right)^{-0.618} + 0.160i\left(\frac{f}{\sigma}\right)^{-0.618}\right\}. \tag{3.9}$$

Here ϕ is the porosity, α_∞ the tortuosity, σ the flow resistivity, f the frequency of sound, and c_0 the speed of sound in air. The characteristic impedance and the complex wavenumber can then be used to calculate the surface impedance and the plane wave, normal incidence absorption coefficient of a hard-backed layer of thickness d according to the following expression:

$$\alpha = 1 - \left|\frac{z_s - 1}{z_s + 1}\right|^2, \tag{3.10}$$

where

$$z_s = z_b \coth(-ik_b\, d). \tag{3.11}$$

Table 3.3 presents a summary of the effective nonacoustical properties for the five types of plants that have been deduced using the adopted equivalent fluid model and proposed optimisation procedure. Table 3.4 gives a summary of these parameters for the two types of soils, and presents a comparison between the directly measured values of the flow resistivity and porosity of the two types of soil and those deduced via the optimisation algorithm.

Figure 3.9 presents the measured absorption coefficient spectra for the five plant species. The grey areas in these five graphs show the variability in the experimental data, which were obtained by collating the results obtained for the five specimens of each plant species. The black lines in these graphs correspond to the absorption coefficient spectra predicted for one representative plant specimen of each species, which is

Table 3.3 The measured and deduced nonacoustical properties for the plants

Plant species	Effective flow resistivity (Pa·s/m²)	Porosity	Effective tortuosity	Effective thickness (m)	Relative error in the absorption coefficient (%)
Geranium zonale – 1	63.28	0.98	2.10	0.14	2.7
Geranium zonale – 2	54.71	0.98	1.49	0.16	0.3
Geranium zonale – 3	35.16	0.98	1.64	0.16	2.5
Geranium zonale – 4	75.16	0.98	1.82	0.13	2.3
Geranium zonale – 5	44.25	0.98	1.37	0.15	0.8
Hedera helix – 1	7.59	0.99	1.00	0.19	0.8
Hedera helix – 2	9.09	0.99	1.02	0.18	0.2
Hedera helix – 3	8.16	0.99	1.12	0.17	0.2
Hedera helix – 4	19.25	0.99	1.44	0.15	1.0
Hedera helix – 5	6.80	0.99	1.00	0.15	2.5
Pieris japonica – 1	14.65	0.98	1.21	0.25	4.0
Pieris japonica – 2	14.00	0.98	1.28	0.33	2.1
Pieris japonica – 3	9.77	0.98	1.35	0.23	2.7
Pieris japonica – 4	21.07	0.96	1.04	0.35	1.2
Pieris japonica – 5	13.55	0.96	1.30	0.28	0.4
SPrimula vulgaris – 1	22.46	0.96	1.12	0.23	0.0
SPrimula vulgaris – 2	16.60	0.96	1.06	0.25	0.5
SPrimula vulgaris – 3	39.78	0.96	1.04	0.28	1.0
SPrimula vulgaris – 4	13.86	0.96	1.28	0.29	1.4
SPrimula vulgaris – 5	15.15	0.96	1.08	0.25	0.6
WPrimula vulgaris – 1	511.95	0.96	2.04	0.08	0.4
WPrimula vulgaris – 2	505.80	0.96	1.85	0.08	1.8
WPrimula vulgaris – 3	318.61	0.90	1.00	0.18	5.6
WPrimula vulgaris – 4	258.30	0.96	1.59	0.11	0.4
WPrimula vulgaris – 5	285.66	0.96	1.74	0.10	0.7

Table 3.4 The measured and deduced nonacoustical properties for soils

Soil type	Flow resistivity (Exp.) (Pa·s/m²)	Flow resistivity (Predicted) (Pa·s/m²)	Porosity (Exp.)	Porosity (Predicted)	Tortuosity (Predicted)	Thickness (m)	Relative error in the absorption coefficient (%)
Substratum	7,600	7,218	0.76	0.76	1.36	0.07	0.1
Clay-based	566,350	620,650	0.39	0.39	2.05	0.10	2.0

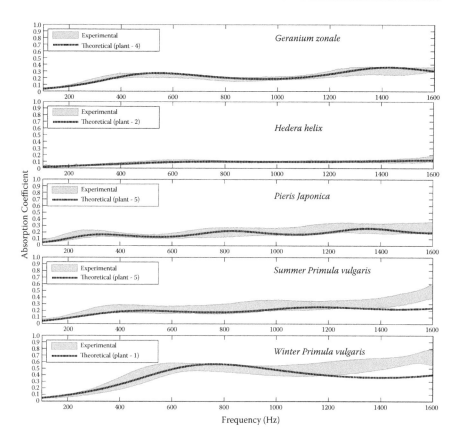

Figure 3.9 The acoustic absorption coefficient data for the five plant specimens of each of the five types of plants, and the absorption coefficient spectra for one particular plant predicted using the equivalent fluid model.

referenced in Table 3.1. Figure 3.10 presents the measured and predicted absorption coefficient spectra for the two types of soils. Table 3.3 and Table 3.4 also present the relative error between predicted and measured absorption coefficient spectra. This mean error in the predicted absorption coefficient was calculated using the following expression:

$$E = \frac{\sum_{n=1}^{N} \left| \alpha^{\mathrm{exp}}(f_n) - \alpha^{th}(f_n, x_0) \right|}{\sum_{n=1}^{N} \alpha^{\mathrm{exp}}(f_n)} \times 100\% .$$

(3.12)

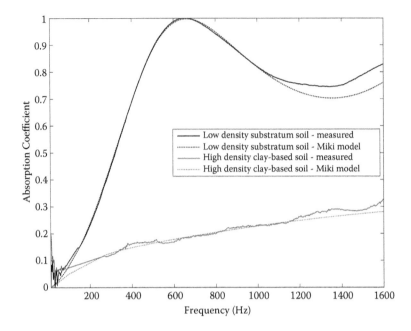

Figure 3.10 Measured and predicted absorption coefficient spectra for the soils using the equivalent fluid model.

The absorption coefficient data presented in Figure 3.9 suggest that plants are able to absorb a considerable proportion of the energy in the incident acoustic wave. In particular, the absorption coefficient of a 100-mm-high specimen of *Winter Primula vulgaris* is able to absorb 60% of the incident acoustic energy in the frequency range between 700 and 900 Hz. The other plants appear to be less efficient absorbers and the percentage of the incident energy absorbed does not exceed 35%. The total leaf area of these plants is smaller than that of *Winter Primula vulgaris*.

The lowest absorption coefficient is found in the case of *Hedera helix* plants, which have the minimum leaf area amongst the other four plants (see Figure 3.9). The highest variability in the standard deviation is found in the *Winter Primula vulgaris* (0.16). This could be related to Primrose plant leaves scattering more acoustic energy compared to plants with a lower leaf surface area, such as *Hedera helix* (0.03). The average area of a single leaf of *Winter Primula vulgaris* is around 0.006 m², whereas the surface area of *Hedera helix* plants is about 10 times lower.

As the frequency of sound increases above 1,200 Hz, the discrepancy between the measured and predicted absorption coefficient spectra also increases, particularly for plants with a relatively high total leaf area. In this frequency regime, the scattering by leaves becomes more pronounced and this effect is impossible to predict accurately with the adopted equivalent fluid model. The relative error in the absorption coefficient is found to be higher in plants with a high total leaf area. The frequency range of the impedance tube needs to be extended farther to be able to measure the relative proportion of the incident sound energy dissipated due to the scattering and viscous/inertial effects. This presents an experimental challenge as the size of the impedance tube needed to accommodate a representative plant specimen has to be relatively large so that the maximum frequency, which can be attained in the ISO 10534-2[1] experiment, remains limited to satisfy the plane wave approximation.

The measured and predicted absorption coefficient spectra shown in Figure 3.9 suggest that the equivalent fluid model approach can provide accurate (with the prediction error of $E < 2\%$, Equation (3.12)) predictions of the acoustic behaviour of soils.

There is a considerable difference in the ability of the two types of soil to absorb sound. A 70-mm-thick layer of light-density soil substrate is able to absorb between 75% and 100% of the incident sound energy in the frequency range above 400 Hz. This acoustic absorption performance is close to that expected from a layer of fibreglass of similar thickness. The results that are presented in the following sections show that the presence of living plants on the top of this type of soil does not result in a substantial increase in the absorption coefficient. The absorption of the 100-mm-thick layer of clay-based soil is limited to 30% of the incident energy across the considered frequency range. The porosity of clay-based soil is relatively small, while the flow resistivity is relatively high. Consequently, there is a large mismatch between the impedance of air and that of soil, which impedes the penetration of the incident sound energy into the porous structure of soil. The results, which are presented in the following sections, show that the presence of living plants on the top of this type of soil can lead to a significant increase in the soil absorption coefficient.

It is now of direct interest to compare the deduced values of the nonacoustical parameters presented in Table 3.3 against some morphological plant characteristics to understand better how these characteristics affect the acoustic absorption by plants. The analysis of the measured absorption coefficient spectra (Figure 3.9) and the plant characteristics listed in Table 3.1 suggests that the ability of plants to absorb sound relates best to the leaf area within the volume occupied by the plant. Figure 3.11 presents the dependence of the effective flow resistivity on the total area of leaves. The data presented in this figure illustrate that the behaviour of the effective flow resistivity follows two distinct patterns. It can be shown that these patterns are controlled by the dominant angle of leaf orientation. Figure 3.12 presents graphically the definition of the dominant angle of leaf orientation that is taken here with respect to the plant stem growing in the direction of the normally incident sound wave.

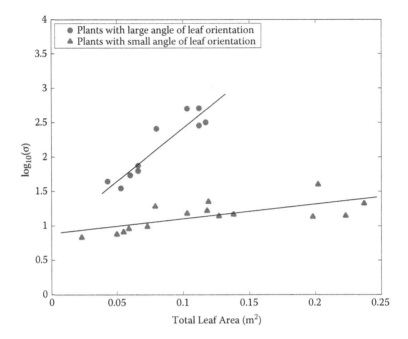

Figure 3.11 Dependence of the flow resistivity on the total area of leaves.

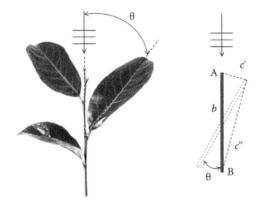

Figure 3.12 Definition of the dominant angle of leaf orientation.

According to the data for the dominant angle of leaf orientation (see Figure 3.12), the investigated plants can be separated into the following two groups: (1) plants in which the leaves are oriented largely in parallel to the direction of the incident sound wave (small angle of leaf orientation), and (2) plants in which the leaves are oriented largely perpendicular to the direction of the incident sound wave (large angle of leaf orientation).

This angle determines the effective path length that the incident acoustic wave follows while propagating through the plant leaf with a characteristic leaf dimension (b) and dominant angle of leaf orientation (θ) as shown in Figure 3.12. Apparently, the effective path length between point A and point B controls the maximum value of the tortuosity that needs to be adopted to minimise the discrepancy between the equivalent fluid model[3] and the experimental data for the acoustic absorption coefficient. It seems sensible to assume that the leaf shown in Figure 3.12 should be centred to stay right in the middle of the path of the incident wave that propagates between some arbitrary points A and B. In this case, the obstruction created by this leaf will be symmetrical so that the effective path length, $c = c' + c''$ (see Figure 3.12, right), propagated by this sound wave either on the left- or on right-hand side of the leaf, will appear identical. It is assumed here that the leaf area density in the considered plants is relatively high. Therefore, the leaves are densely arranged so that the height of the plant and length of the leaf stems have little or no effect on the maximum value of the tortuosity. In this case, it is easy to show that the effective path length for the acoustic wave propagating through the plant is controlled by the leaf angle only and is given by:

$$c = b\left(\cos\frac{\theta}{2} + 2\sin\frac{\theta}{2} \right). \tag{3.13}$$

Following the definition of the structure factor by Zwikker and Kosten,[10] the tortuosity can be expressed as:

$$\alpha_\infty = c / b = \cos\frac{\theta}{2} + 2\sin\frac{\theta}{2}. \tag{3.14}$$

This implies that the maximum theoretical value of the tortuosity for a plant with leaves oriented perpendicular to the direction of wave propagation, i.e., $\theta = \dfrac{\pi}{2}$, is $\alpha_\infty = 3 / \sqrt{2} \simeq 2.12$.

The effect of the dominant angle of leaf orientation, i.e., effective tortuosity, is particularly visible in Figure 3.13, which presents the flow resistivity as a function of the ratio of the leaf area to the volume occupied by the plant (leaf area density). These data enable us to propose two semiempirical relations between the effective flow resistivity of a plant and its leaf area density:

$$\log_{10}\sigma = 0.0083 A_v + 1.413, \text{ for } \theta > 70°; \tag{3.15}$$

$$\log_{10}\sigma = 0.0067 A_v + 0.746, \text{ for } \theta > 40°. \tag{3.16}$$

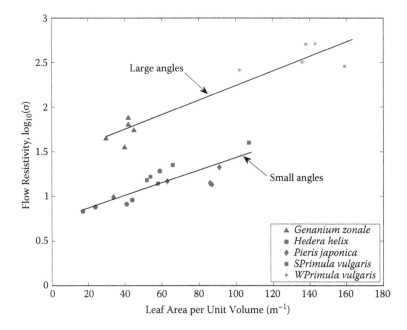

Figure 3.13 Flow resistivity as a function of the leaf area density.

These relations are exponential. Exponential relations between the light reflectivity and leaf area index have been observed in the past by researchers working on remote sensing of biomass.[11] Relations (3.15) and (3.16) are different to those reported in the work by Taherzadeh et al.[12] for plants with much lower leaf area density. There is a considerable difference in the values of the effective flow resistivity between different plants (see Table 3.3). The maximum value of the flow resistivity is for *Winter Primula vulgaris*, specimen 1 (511 Pa·s·m^{-2}). The minimum value of the flow resistivity is for *Hedera helix*, specimen 5 (6.8 Pa·s·m^{-2}). These two plant specimens are characterised by the largest (143 m^{-1}) and smallest (17 m^{-1}) leaf area density, respectively. In terms of the absorption coefficient spectra, these plants exhibit the highest and lowest absorption coefficient spectra, respectively.

3.7 ABSORPTION OF SOIL IN THE PRESENCE OF A PLANT

The measured and deduced nonacoustical characteristics of plants and soils can now be used to predict the acoustic absorption coefficient of a layer of soil on which a plant specimen was placed. This can be achieved using a standard transfer matrix approach, which was discussed in the previous sections of this chapter. In this experiment, the plant specimen was not planted into the soil, but simply placed on the top layer of soil so that the porous structure of soil was not disturbed.

A comparison between the predicted absorption coefficient for the soil layer only and soil and plant system, and the measured absorption coefficient is presented in Figure 3.14 for the two types of soil and five types of plants studied in this work. The individual plant specimens used in the calculations for each species are indicated in the captions of the graphs and the appropriate nonacoustical parameters for these specimens are in Table 3.3. The texture of the two soils determines the pore size distribution that controls their acoustical properties. The presence of fibres, large particles of perlite, and polymer gel in the substratum gives rise to larger pores and, therefore, have a major influence on its acoustic absorption coefficient (see Figure 3.14), whereas the texture of the high-density, clay-based soil is finer with closely arranged particles that are less than 2 mm in diameter.

These results show that different plant species have markedly different effect on the absorption coefficient of soil. The presence of a plant on the top of heavy density, clay-based soil results in a noticeable increase in the absorption coefficient across a broad frequency range (see Figure 3.14). The greatest increase (up to 50%) is observed in the case of the plants with the largest leaf area density, i.e., when *Winter Primula vulgaris* is planted. On the contrary, plants with a relatively small leaf area density have little or no effect on the measured absorption coefficient of this type of soil, i.e., *Hedera helix* or *Pieris japonica*. The adopted equivalent fluid model is able to predict the absorption coefficient spectrum of the plant–soil system with an error of ±8%. The most accurate predictions are attained in the case of *Summer Primula vulgaris*, *Pieris japonica*, and *Geranium zonale* (error less than 3.5%) for both soil types. The least accurate predictions are attained in the case of *Winter Primula vulgaris* (±4.8%) and *Hedera helix* (±8%). This can be attributed to a larger variability in the experimentally determined absorption coefficient of *Winter Primula vulgaris*, which is illustrated in Figure 3.8. The absorption coefficient of *Winter Primula vulgaris* seems sensitive to changes in the dominant angle of leaf orientation and plant arrangement, which was achieved in the experiment when this plant was placed on the top of the layer of clay-based soil. These may not be identical in the plant only and plant plus soil experiments. In the case of *Hedera helix*, the model consistently overpredicts the absorption coefficient of soil in the presence of the plant. The reduction in the soil absorption coefficient in the presence of *Hedera helix* also has been observed in other experiments.[13] It can be argued that it is rather difficult to make an accurate prediction for *Hedera helix* plants because they have a low area per unit volume and low weight of leaves.

The presence of *Geranium zonale*, *Pieris japonica*, or *Summer Primula vulgaris* on the top of the low-density soil substrate has a relatively small

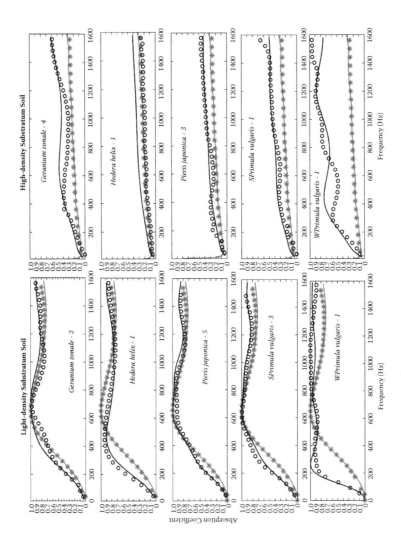

Figure 3. 14 Measured and predicted absorption coefficient spectra for the five plant specimens with low-density substratum soil and high density clay-based soil using the transfer matrix model. Measured data for soil plus plant (o), measured data for soil only (*), predicted for soil plus plant (solid black line), and predicted for soil only (solid grey line).

effect on the soil absorption coefficient (see Figure 3.14, left). The presence of *Winter Primula vulgaris* results in up to an 80% increase in the soil absorption coefficient at frequencies below 400 Hz. There is a 15–20% increase in the soil absorption coefficient at frequencies below 400 Hz and 10–15% increase at frequencies above 800 Hz when *Winter Primula vulgaris* is added to light-density soil substrate. Generally, there is a good agreement between the modelled and predicted acoustic absorption spectra with a maximum error of 8% in the case of *Hedera helix* plants placed on top of the clay-based soil.

3.8 MODELLING THE RANDOM INCIDENCE ABSORPTION COEFFICIENT OF SOIL WITH AND WITHOUT PLANTS

In the previous sections of this chapter, it was shown that it is possible to predict the plane wave, normal incidence acoustic absorption coefficient of soil using a model that treats soil as an equivalent fluid medium. It also was shown that the same model can be used to predict the normal incidence acoustical absorption coefficient of plants. The acoustical absorption of a plant–soil system was determined using a transfer matrix approach provided the characteristic impedance and propagation constant for each of the two equivalent layers in this system are known. The use of the equivalent fluid model enables us to account for the viscous losses and thermal dissipation of the sound wave as it propagates through the plant foliage and through the layer of porous soil. It is believed that these mechanisms can explain the acoustic absorption coefficient spectra observed in the impedance tube experiments with the plants.

In this section, we extend the same equivalent fluid model to simulate the random incidence absorption coefficient. We will use Equation (3.8) and Equation (3.9) to predict the angular-dependent absorption coefficient for a hard-backed porous specimen. In this case, we assume that the absorption coefficient of soil or plant can be angular-dependent so that Equation (3.10) and Equation (3.11) then become:

$$\alpha(\varphi) = 1 - \left| \frac{z_s(\varphi) - 1}{z_s(\varphi) - 1} \right|^2 \qquad (3.17)$$

and

$$z_s(\varphi) = z_b \, / \cos\varphi \coth(-ik_b d \cos\varphi), \qquad (3.18)$$

respectively. Here, φ is the angle of incidence.

A simple way to relate the random incidence absorption coefficient with its plane wave counterpart is to use the formula suggested by London in his 1950 paper.[14] A common expression for the random incidence absorption coefficient is:

$$\bar{\alpha} = 2 \int_{0}^{\pi/2} \alpha(\varphi) \cos\varphi \, \sin\varphi \, d\varphi. \tag{3.19}$$

London showed that the random incidence absorption coefficient can be approximated by either of the following two expressions (cf. ISO 10534-2,[15] Equations (3.11) and (3.14)):

$$\bar{\alpha} = \frac{4}{Ze}\left[\text{in}(1+Z_e) - \frac{Z_e}{1+Z_e} \right] \text{ or } \bar{\alpha} = \frac{8}{Z_e^2}\left[1+Z_e - \frac{1}{1+Z_e} - 2\text{in}(1+Z_e) \right], \tag{3.20}$$

where

$$Z_e = \frac{1+\sqrt{1-\alpha(0)}}{1-\sqrt{1-\alpha(0)}} \tag{3.21}$$

is the so-called equivalent impedance.

The soil used in this study was a light-density soil substratum whose basic characteristics were discussed in the previous sections of this chapter. The effect of soil compaction when it is laid in the reverberation chamber and the effect of added water to this soil substratum are rather complex and seem to be confined to the top layer of soil. Therefore, it is of direct interest to determine the effect of moisture on the random incidence absorption coefficient of soil. For this purpose 10 m^2 of a 100-mm-thick layer of Canevaflor soil substrate was laid in the reverberation chamber at the University of Sheffield. The testing procedure was carried out in accordance with the ISO 354.[15] Controlled amounts of water were added to the porous soil to simulate a range of water saturation conditions. Figure 3.15 shows the measured random incidence absorption coefficient of soil for three levels of water saturation: $S_r = 0$, 40, and 100%.

Soil substrate used in this work consists of pores of a relatively large diameter (200 øm and above). It also contains a proportion of meso-pores (>1 mm scale) that are created around the particles of perlite and polymer gel, which are relatively large. If the top surface of this material is sprayed with water, then the water tends to penetrate into the first few millimetres of the soil sample just to be absorbed by the polymer gel particles confined to the top. This phenomenon results in a material with pore size stratification. The acoustical properties of this multilayered

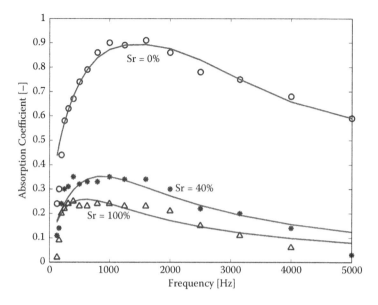

Figure 3.15 The measured (markers) and modelled (solid lines) random incidence absorption coefficient of a 100-mm-thick layer of soil substrate at different degrees of saturation.

material can no longer be modelled using a uniform pore size distribution assumption and pore size stratification information is required. A relatively straightforward method that was used to account for this stratification is Horoshenkov et al.'s two-layer model.[6] In this way, the acoustical properties of the top, water-saturated layer and the bottom, relatively dry layer of soil can be modelled using two sets of nonacoustical parameters. Figure 3.15 shows the random incidence absorption coefficient predicted for these three water saturation levels using the equivalent fluid model by Miki[3] and two-layer transfer matrix approach. London's formulae (Equation (3.20)) were employed to estimate the random incidence absorption coefficient from its normal incidence, plane wave counterparts predicted by Equation (3.19) for $\phi = 0$. Table 3.5 presents the summary of the nonacoustical parameters used for modelling the observed acoustic absorption coefficient behaviour.

The work by Horoshenkov et al.[6] shows that the absorption coefficient of soil can be enhanced by adding a plant with a relatively large leaf area density and angle of leaf orientation. In this work, the greatest increase (up to 50%) was observed in the case when *Winter Primula vulgaris* or *Geranium zonale* were planted in high-density soil. Light-density, high-permeability soils, such as soil supplied by Canevaflor, exhibit high

Table 3.5 The nonacoustical parameters used in the modelling

Degree of saturation (%)	Layer	Flow resistivity (Pa s m^{-2})	Porosity	Tortuosity	Layer thickness (mm)
0	top	7.65 × 10^3	0.25	1.0	16.0
0	bottom	1.11 × 10^5	1.0	1.0	84.0
40	top	1.62 × 10^6	0.01	1.0	3.4
40	bottom	8.75 × 10^5	1.0	1.0	96.6
100	top	1.48 × 10^6	0.01	1.0	6.0
100	bottom	2.49 × 10^6	1.0	1.0	94.0

values of acoustic absorption that are comparable with that expected from a layer of fibreglass of similar thickness. The presence of a plant on this type of soil can have a significant effect on the plane wave, normal incidence absorption coefficient spectra in the low frequency range. It is now of interest to study the random incidence absorption coefficient of soil with and without plants. For this purpose, we can use the nonacoustical characteristics for some plants studied by Horoshenkov et al.[6] together with the characteristics for soil derived in the previous section of this chapter. We will adopt the same equivalent fluid model by Miki[3] that will be used to predict the acoustical properties of the following three-layer system: plant layer, top soil layer, and bottom soil layer.

Figure 3.16 presents the random incidence absorption coefficient of soil with and without a plant. The soil parameters used in this simulation correspond to the dry soil specimen as shown in Table 3.5. Two different plants were considered here: *Winter Primula vulgaris* (winter primrose) and *Hedera helix* (green ivy). These plants are characterized by markedly different leaf area densities. The leaf area density of the *Winter Primula vulgaris* specimen used in this calculation was 141 m^{-1}, whereas this parameter was 41 m^{-1} in the case of the *Hedera helix* specimen. The other nonacoustical parameters of plants used in this simulation are shown in Table 3.6. These characteristics were reproduced from Table 3.3 for *Hedera helix-2* and *Winter Primula vulgaris-1* specimens.

It is clear from the results shown in Figure 3.16 that the presence of a relatively small plant has a noticeable effect on the random incidence absorption coefficient of soil. In the case of a plant with a relatively small leaf area density (i.e., *Hedera helix*), the increase in the absorption coefficient of soil is relatively small (below 13%) and it is mainly confined to the higher frequency range above 1,000 Hz. In the case of a plant with much larger lead area density (i.e., *Winter Primula vulgaris*), this increase is much more pronounced below 1,000 Hz (up to 15% at 600 Hz) and above 1,000 Hz (up to 35% at 4,000 Hz).

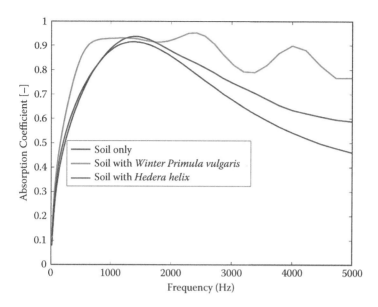

Figure 3.16 The simulated random incidence absorption coefficient of dry soil with and without a plant.

Table 3.6 The nonacoustical parameters used in the modelling

Plant	Flow resistivity (Pa s m^{-2})	Porosity	Tortuosity (α_∞)	Plant height (mm)
Winter Primula vulgaris	512	0.96	2.05	80
Hedera helix	9	0.99	1.1	180

3.9 CONCLUSIONS

This chapter illustrates that some soils can exhibit very high values of the acoustic absorption coefficients. The acoustic absorption of high density soils can be strongly affected by the presence of moisture. This effect is observed across a broad range of frequencies of interests. The acoustic absorption of low-density soils does appear to be affected significantly by the presence of moisture, except in a limited part of the higher frequency range. The acoustic absorption coefficient of soils can be further enhanced by carefully selected types of low-growing plants. The increase in the absorption coefficient of a soil–plant system is particularly pronounced in the case of high-density, low-porosity soils. The acoustical properties of low-growing plants can be largely explained by the leaf area density and dominant angle of leaf orientation. These two parameters can be related to the effective flow resistivity and tortuosity in an equivalent fluid model that can be adopted to explain the observed acoustic absorption behaviour of plants. It has been shown that the random incidence absorption coefficient of soil with plants can be modelled using the approach proposed by London.[14,15]

REFERENCES

1. ISO. 1998. Acoustics—Determination of sound absorption coefficient and impedance in impedance tubes. Part 2: Transfer-function method (ISO 10534-2). Geneva, Switzerland: International Standards Organization.
2. Horoshenkov, K. V., and M. H. Mohamed. 2006. Experimental investigation of the effects of water saturation on the acoustic admittance of sandy soils. *The Journal of the Acoustical Society of America* 120: 1910.
3. Miki, Y. 1990. Acoustic properties of porous materials – Generalization of empirical models. *Journal of the Acoustical Society of Japan* (E) **11** (25–28).
4. Horoshenkov, K., and M. Swift. 2001. The effect of consolidation on the acoustic properties of loose rubber granulates. *Applied Acoustics* 62 (6): 665–690.
5. Horoshenkov, K., K. Attenborough, and S. Chandler-Wilde. 1998. Padé approximants for the acoustical properties of rigid frame porous media with pore size distributions. *The Journal of the Acoustical Society of America* 104: 1198.
6. Horoshenkov, K. V., A. Khan, and H. Benkreira. 2013. Acoustic properties of low growing plants. *Journal of the Acoustical Society of America* 133 (5): 2554–2565.
7. Leclaire, P., O. Umnova, K. V. Horoshenkov, and L. Maillet. 2003. Porosity measurement by comparison of air volumes. *Review of Scientific Instruments* 74 (3): 1366–1370.
8. ISO. 1999. Acoustics—Materials for acoustic applications: Determination of airflow resistance (ISO 9053:1991). Geneva, Switzerland: International Standards Organization.
9. Nelder, J. A., and R. Mead. 1965. A simplex method for function minimization. *The Computer Journal* 7 (4): 308313.
10. Zwikker, C., and C. Kosten. 1949. *Sound absorbing materials*. Amsterdam: Elsevier.
11. Heiskanen, J. 2006. Estimating above ground biomass and leaf area index in mountain birch forest using ASTER satellite data. *International Journal of Remote Sensing* 27: 1135–1158.
12. Taherzadeh, S., K. Attenborough, and I. Bashir. 2012. Attenuation of sound by crops and hedges. Paper presented at the Proceedings Acoustics 2012, Société Française d'Acoustique, Nantes, France.
13. Horoshenkov, K., and A. Khan. 2012. Acoustic shielding effect of plants and hedges. (FP7 HOSANNA Deliverable 2.3). Online at: http://www.greener-cities.eu/

14. London, A. 1950. The determination of reverberant sound absorption coefficients from acoustic impedance measurements. *Journal of the Acoustical Society of America* 22: 263–269.
15. ISO. 2003. Acoustics—Measurement of sound absorption in a reverberation room (ISO 354:2003). Geneva, Switzerland: International Standards Organization.

Chapter 4

Acoustical characteristics of trees, shrubs, and hedges

*Timothy Van Renterghem, Dick Botteldooren,
Jian Kang, Kirill Horoshenkov, and Hong-Seok Yang*

Abstract. This chapter discusses and illustrates the basic interactions between sound waves and vegetation. Only absorption by plant elements, as a result of thermo-viscous effects at their surfaces and by damped vibrations, leads to effective loss of acoustical energy in a sound wave. Scattering of sound, by diverging sound energy away from the direct path between a source and receiver, can be efficient as well to reduce noise levels at a single receiver. Reflection and diffraction at vegetation is considered to be of minor importance in typical road traffic noise applications. These basic interactions are illustrated by impedance tube measurements, leaf vibration measurements, standardized scattering measurements in a reverberant room, *in situ* scattering experiments near trees, and by numerical calculations.

4.1 INTRODUCTION

The interaction between sound waves and vegetation can be categorized in either direct acoustical effects or indirect acoustical effects. Direct effects (see Figure 4.1) are linked to two main processes, namely, redistribution of sound energy on the one hand, and absorption on the other hand. Redistribution of sound occurs by reflection, diffraction, and scattering when interacting with plant elements, such as trunks, branches, twigs, and leaves. Only absorption leads to effective loss of acoustic energy, which actually means transfer to heat. Nevertheless, redistribution of sound energy can be effective to achieve noise reduction at a single receiver as well. In this chapter, measurements and numerical simulations illustrating such direct interactions are discussed.

Indirect acoustic effects are caused by a secondary action because of the presence of vegetation. However, this does not mean that these effects are less important than the direct effects, and often they outperform such direct effects. The development of an acoustically soft soil underneath vegetation, called a *forest floor,* is the result of plant rooting and the formation of a humus layer consisting of plant litter. Also, the change of the

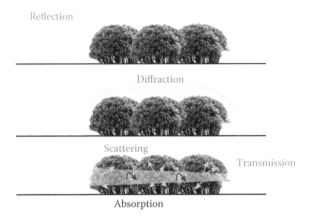

Figure 4.1 Illustration of the direct interactions between sound waves and vegetation.

microclimatology by canopies, in turn, influencing the properties of the sound propagation medium, can be an important indirect effect.

4.2 ABSORPTION BY LEAVES

Two physical phenomena can lead to absorption of sound. Firstly, at the interface between the surface of a leaf and air, absorption is caused by thermo-viscous effects. Secondly, the sound waves lose part of their energy by making the leaves vibrate; the leaves follow the oscillations of the incident waves. Part of the energy of the vibrating leaves is reemitted and this energy then again becomes airborne sound. Another part of these vibrations will be damped, because leaves are connected to other parts of the plant that are more rigid. In this way, part of the energy of the sound wave is effectively lost by transfer to heat.

4.2.1 Measuring leaf absorption

The reflections of sound from different types of leaves of low-growing plants have been measured in an impedance tube (Figure 4.2(a)–(d)). This is the traditional way of measuring the acoustical absorption coefficient of a material under controlled conditions in the laboratory. As with many materials, there is an increase in the absorption abilities of leaves (and other plant structures) with increasing sound frequency. At low frequencies, the absorption coefficient is close to zero. In such an experiment, all absorption mechanisms are observed simultaneously and thermo-viscous- and vibration-related effects cannot be singled out. Therefore, their relative importance is difficult to assess.

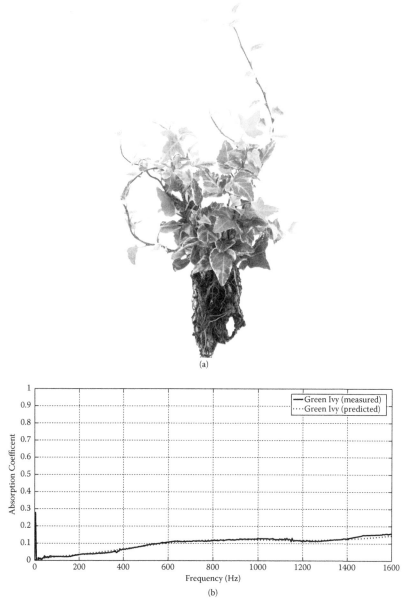

(a)

(b)

Figure 4.2 Absorption measurements (in an impedance tube) of leaves (in absence of soil) of low-growing plants. The differences in absorption among different species can be large. Ivy (a,b) is an example of a low-absorption plant, while Primrose leaves (c,d) lead to rather strong absorption.[1] Basic plant characteristics allow accurate prediction of the dependence of absorption on sound frequency for each plant.

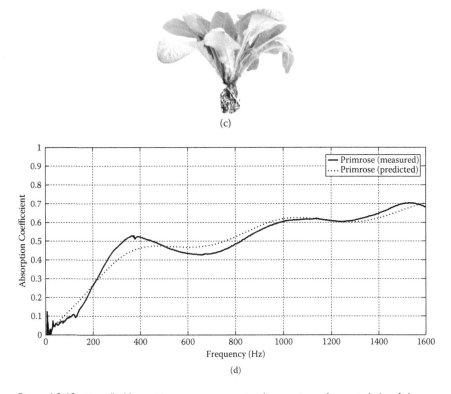

(c)

(d)

Figure 4.2 (Continued) Absorption measurements (in an impedance tube) of leaves (in absence of soil) of low-growing plants. The differences in absorption among different species can be large. Ivy (a,b) is an example of a low-absorption plant, while Primrose leaves (c,d) lead to rather strong absorption.[1] Basic plant characteristics allow accurate prediction of the dependence of absorption on sound frequency for each plant.

The measurements show that there are large differences among plant species. Some types of plants hardly absorb sound, while for others, effects can be quite pronounced (see examples in Figure 4.2(b)(d)). The parameters that influence sound absorption have been identified. Key parameters are the leaf area density (the amount of surface presented by all leaves in a given plant volume) and the dominant angle of the leaves relative to the angle of incidence of the sound. The leaf area density is strongly related to the viscous drag experienced by air moving through the plant, which is measured as flow resistivity. In general, a high flow resistivity prevents sound waves from penetrating a material; the latter, however, is essential to obtain absorption of acoustical energy. The dominant angle of the leaves defines the tortuosity, which can be seen as the detour a sound wave needs to make when travelling through a material, relative to a straight-line propagation. Absorption by plants is enhanced when the sound is incident at right angles to the surface of the leaves. The frequency-dependent absorption by leaves can be well predicted with reasonably simple numerical models.[1]

4.2.2 Measuring leaf vibrations

Leaf vibrations have been measured in an anechoic chamber (Figure 4.3). A single leaf was positioned in front of a loudspeaker, and a laser Doppler vibrometer (LDV) was directed on a spot of the leaf. In this way, the vibrations of the leaf under study could be tracked.

These measurements show that leaves vibrate with the incoming sound, exhibiting a pressure variation over time similar to that of the incoming airborne sound wave. Note that these oscillations are too rapid to be seen by the human eye.

The pressure difference measured over the leaf (at about 1 cm in front and 1 cm at the back from the leaf plane) suggests that leaf vibrations could contribute to noise reduction. However, this is a high-frequency effect typically near 4–5 kHz and is strongly related to the size of the leaves. Smaller leaves are expected to exhibit such effects at even higher sound frequencies; for larger leaves, this is expected at somewhat lower frequencies. The irregular form of the measured curve shows the complex interactions and resonances in the leaf associated, for example, with the presence of bending waves in the non-uniform leaf plane. Nevertheless, this general behaviour can be reasonably well modelled numerically by assuming uniform leaf properties (Figure 4.4).

Figure 4.3 Experimental setup for measuring the vibrations of a single leaf in a sound field with a laser Doppler vibrometer, in combination with measuring the acoustic pressure drop over the leaf.

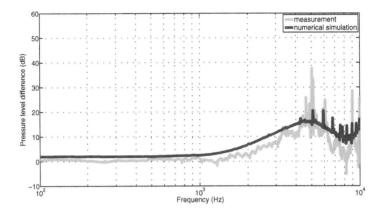

Figure 4.4 Measured and simulated (with a full wave- numerical model) pressure drop over a single *Prunus laurocerasus* leaf.

4.3 REFLECTION AND DIFFRACTION BY VEGETATION

Reflection of sound will occur at the interface between a vegetated and nonvegetated zone as long as there is a sufficient change in the propagation properties. Additional reflections might come from deeper inside a vegetation belt as well. Tree trunks are considered to be the main actors in the reflection of road traffic noise. With increasing frequency and tree density, reflections become more prominent. Sound energy that is reflected back to the source will not reach a receiver positioned behind the vegetation.

Measurements by Wunderli and Salomons[2] showed that, at relevant frequencies in road traffic noise, the reflections are about 10 dB lower than direct sound at the edge of a forest (average trunk diameter equal to 27 cm, for a trunk basal area of 0.5%). This means a relatively small contribution from the reflections to the total sound pressure level at the source side. In some cases, like, for example, when dealing with explosions, such reflections could be relevant because they appear separated in time and can still be high in amplitude (leading to audible echoes). However, road traffic is a continuous source and such reflections will neither be observable nor lead to increased levels.

Diffraction is a term for the physical process of bending of sound around objects. Diffraction is the reason why it does not become fully silent behind a noise wall, even when the source is not visible from the viewpoint of the receiver. Diffraction is most pronounced near the edges of solid objects. Consequently, strong diffraction effects are not expected near a vegetated zone. More specifically, the sound energy transmitted through a vegetated zone will be more important than part of the sound energy diffracted over the top or diffracted along its sides.

4.4 SCATTERING BY VEGETATION

In vegetation, multiple scattering occurs. Sound incident on a trunk, twig, or leaf will change its propagation direction, and is then scattered again by nearby vegetation elements. As a result, part of the sound energy will continuously leave the direct path between source and receiver during propagation through a zone of vegetation, yielding lower sound pressure levels at that receiver.

Scattering leads to the redirection of sound energy, in contrast to absorption where sound energy is effectively lost by conversion to heat. However, the main part of the sound energy will keep on following the direct path between source and receiver. In case of a large number of interactions, for example, for long distance propagation through the vegetation, positive effects can be obtained relative to sound propagation in the absence of vegetation. This is not only due to redirection of sound energy; multiple scattering also enables more absorption.

Scattering will become pronounced when the dimensions of the objects are close to the wavelength of the sound. High sound frequencies mean small wavelengths, low frequencies mean large wavelengths. Given the typical dimensions of the elements in canopies, foliage scattering is a high frequent effect and will become prominent typically above 2 kHz. Also in-between trunks, multiple scattering will be present. The latter effect occurs at much lower frequencies since bigger structures are involved. This means that scattering by trunks, affecting sound frequencies closer to the maxima in typical spectra, could influence total levels effectively. For this reason, a strong emphasis will be put on studying the shielding provided by tree belts in Chapter 5.[3]

4.4.1 Measuring scattering by a pile of leaves in the laboratory

The scattering coefficient of a pile of leaves has been measured in a reverberant chamber[4] using a turning table (Figure 4.5), following a standardised measurement procedure.[5] The scattering coefficient is a value between 0 and 1, and indicates the fraction of the acoustical energy reflected in a nonspecular way. A value of 0 means no scattering at all and indicates a similar behaviour to that of a totally flat surface. A value of 1 means that all reflections from a surface can be considered as scattered. The use of such a specific experimental setup is needed because

Figure 4.5 A turning table setup to measure the scattering coefficient of a bunch of leaves.

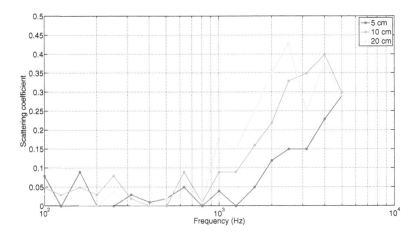

Figure 4.6 Scattering coefficient of a pile of leaves in function of frequency.

scattering properties might strongly depend on the angle of incidence of the sound, and well-thought-out averaging is needed to achieve meaningful results.

This experiment clearly shows that foliage scattering is a high-frequency effect. Below 1 kHz, no relevant effects are observed, as depicted in Figure 4.6. Consequently, foliage will only affect part of the road traffic noise spectrum containing less energy. The scattering properties also depend on the degree of vegetation cover and the thickness of the vegetation layer.

4.4.2 Scattering by a single tree

Scattering by branches, twigs, and leaves in the crown of a tree was measured and found to be pronounced (Figure 4.7). Due to multiple scattering processes, sound waves can have a rather long residence time inside the canopy, and sound only gradually leaves the canopy. Due to this prolonged propagation path, there will be an increase in geometrical spreading of sound and more interaction (e.g., absorption) with the canopy elements. Reverberation time, which is a term typically used in room acoustics, can give an estimation of the magnitude of the multiple scattering process. The larger this value, the longer the residence time of the sound in the canopy and the larger the influence a tree exhibits on the sound field.

A large number of *in situ* measurements were performed. With increasing crown size and sound frequency, multiple scattering inside the crown becomes more pronounced. Even without foliage, a tree is already a good scattering object. In leaves, the reverberation time increased, especially above 1 kHz (Figure 4.8), which is consistent with the previously mentioned laboratory measurements.

Depending on the application, scattering by a single tree can be either wanted or unwanted. In an urban environment, multiple reflections between building facades may strongly amplify sound. The resulting reverberation is known as the (acoustic) *street canyon effect*. Trees can decrease

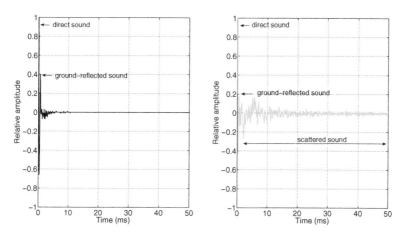

Figure 4.7 Time-domain analysis of pulse propagation near a big tree (b), with the sound source and receiver at the opposite sides of the tree. For comparison, the same results are shown in case of sound propagation in an open field (a).

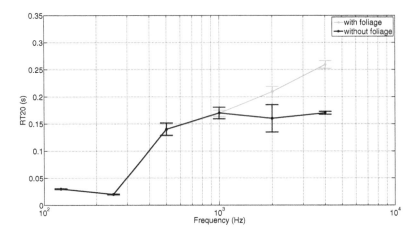

Figure 4.8 Reverberation time, as a quantitative parameter for the degree of scattering of sound in the canopy of a tree, with and without foliage, in function of sound frequency.

reverberation by redirecting part of the sound to leave the street upwardly after interactions with canopy elements, thereby not contributing anymore to this street reverberation. Direct sound propagation between a road traffic noise source and higher stories of buildings will be partly redirected, so that less sound will arrive at these facades. On the other hand, trees in streets might lead to slightly increased sound pressure levels due to downward scattering when both the source and the receiver are below the canopy bottom. Clearly, this is unwanted. A similar effect can be observed near a noise wall with a row of trees behind it. However, due to the high-frequency nature of this downward scattering into the acoustic shadow zone, this was experimentally shown to be a minor problem in road traffic noise applications. The positive effects related to the windbreaking function of such a row of trees behind a noise wall are much larger.[6] This topic is treated in greater detail by Defrance et al.[7] and in Chapter 2, and Van Renterghem et al.[3] and in Chapter 5.

4.4.3 Visualising scattering in the multiple layers in a vegetation belt

A numerical technique that takes into account the curvature of the wavefronts (known as *full-wave* methods) enables visualisation of the multiple scattering processes in a vegetation belt. Scattering occurs at understory vegetation, in the trunk layer, and also by the canopies. With this type of model,[8] the volumes occupied by the trunks can be explicitly included and the influence of trunk positioning can be analysed in detail, as discussed in

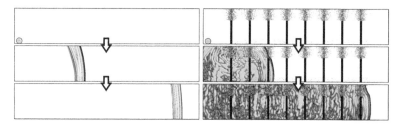

Figure 4.9 Snapshots of the simulated sound field distribution at three moments during propagation through an open field (left) and a vegetation belt (right) obtained using a 3D finite-difference, time-domain calculation.[8,9] An acoustic pulse is initially excited (upper row) and the sound field development (wavefront spreading) during propagation is shown (middle and bottom row). The colour scale is arbitrary: orange and yellow indicate zones of high sound pressure levels, green intermediate levels, and blue low levels. The multiple scattering processes in the different layers of a vegetation belt are clearly visible in the right-hand diagrams.

detail in Chapter 5.[3] The scattering by smaller elements in the canopy and shrub layer are mimicked by using a distribution of scattering elements, with a similar aboveground biomass to that in real vegetation.[10] Figure 4.9 clearly shows the buildup of a highly diffuse sound field inside a vegetation belt, opposed to the cylindrical radiating pattern in case of a road in the absence of a vegetation belt.

REFERENCES

1. Horoshenkov, K. V., A. Khan, and H. Benkreira. 2013. Acoustic properties of low growing plants. *Journal of the Acoustical Society of America* 133(5): 2554–2565.
2. Wunderli, J. M., and E. M. Salomons. 2009. A model to predict the sound reflection from forests. *Acta Acustica United with Acustica* 95(1): 76–85.
3. Van Renterghem, T., K. Attenborough, and P. Jean. 2014. Designing vegetation and tree belts along roads. In *Environmental methods for transport noise reduction*, eds. M. E. Nilsson, R. Klæboe, and J. Bengtsson (pp. 93–119). Oxford, U.K.: Spon Press.
4. Yang, H.-S., J. Kang, and C. Cheal. 2013. Random-incidence absorption and scattering coefficients of vegetation. *Acta Acustica United with Acustica* 99(3): 379–388.
5. *Acoustics-Measurement of Random-Incident Absorption and Scattering Coefficients of Vegetation*. 2004. ISO 17497–1 (ISO, Geneva, Switzerland.).
6. Van Renterghem, T., and D. Botteldooren. 2002. Effect of a row of trees behind noise barriers in wind. *Acta Acustica United with Acustica* 88(6): 869–878.

7. Defrance, J., P. Jean, F. Koussa, T. Van Renterghem, J. Kang, and J. Smyrnowa. 2014. Innovative barriers. In *Environmental methods for transport noise reduction*, eds. M. E. Nilsson, R. Klæboe, and J. Bengtsson (pp. 19–47). Oxford, U.K.: Spon Press.

8. Botteldooren, D. 1995. Finite-difference time-domain simulation of low-frequency room acoustic problems. *Journal of the Acoustical Society of America* 98(6): 3302–3308.

9. Van Renterghem, T., and D. Botteldooren. 2003. Numerical simulation of the effect of trees on downwind noise barrier performance. *Acta Acustica United with Acustica* 89(5): 764–778.

10. Van Renterghem, T., D. Botteldooren, and K. Verheyen. 2012. Road traffic noise shielding by vegetation belts of limited depth. *Journal of Sound and Vibration* 331(10): 2404–2425.

Designing vegetation and tree belts along roads

Timothy Van Renterghem,
Keith Attenborough, and Philippe Jean

Abstract. This chapter provides design rules for tree/vegetation belts along roads. The basic interactions between sound waves and the components of a tree/vegetation belt (soil, shrubs/hedges, trunks, and canopies) are discussed, and the parameters of importance identified. The use of an optimized planting scheme could turn a tree belt along a road into an efficient noise-reducing measure. While, in general, one should strive for rather high biomass densities to obtain useful noise reduction, some interesting approaches are presented to relax such demands without affecting noise shielding too much. A strip of trees also has an effect on the stability of the lower part of the atmospheric boundary layer, and, thus, on its refractive properties. As a consequence, an additional positive effect is expected during the night by the presence of a tree belt, which outperforms the slightly worsened meteorological situation during daytime. A topic of great concern is the strong reduction in efficiency of a noise wall under downwind sound propagation conditions. A row of trees, acting as a windbreak, can recover at least part of the shielding that was lost by the action of the wind. Design rules are provided that are applicable for scenarios with single noise walls, noise walls on either side of the road, and in case of steep berms. Downwind receivers at close distance and farther away are considered.

5.1 INTRODUCTION

A tree belt is an interesting solution to achieve road traffic noise reduction. Noise shielding is obtained as a combination of multiple scattering by mainly the tree trunk layer (direct effect) and the presence of a forest floor (indirect effect). In Section 5.2, guidelines are provided to optimise noise shielding by non-deep tree belts along roads. Non-deep belts require less land and can be easier to apply. Predicted noise reductions in a typical road traffic situation are presented.

In Section 5.3, the possibility to improve wind fields by means of rows of trees near noise-reducing devices like noise walls and earth berms under downwind conditions is discussed (see also this volume, Chapter 2[1]). This is a clear example of an indirect effect that can be exploited to decrease noise levels. A number of design guidelines for the application of a row of trees as a windbreak are provided. In addition, it is shown how a tree belt influences temperature-related refraction effects during day and night.

5.2 DESIGNING VEGETATION BELTS NEAR ROADS

5.2.1 Introduction and research methodology

In this section, some important parameters determining the noise reduction by tree belts are identified.

Planting a tree belt of finite width and depth along roads can be an interesting way to abate road traffic noise. Hitherto published findings in scientific and non-scientific literature related to the effectiveness of vegetation/tree belts are divergent. Some conclude that there are almost no effects, while others claim strong acoustical benefits. At least, this illustrates that achieving significant noise abatement with vegetation is not trivial.

Given the nature of this type of research, design rules are not derived from field studies, but are mainly based on numerical simulations. Although measurements are always useful for validation purposes, for the specific case of tree or vegetation belts, there are huge practical difficulties with experimental setups: it takes many years for trees to develop, and trees cannot be easily rearranged to test specific planting schemes. Measuring the properties of existing tree belts along roads is of limited use, because it deviates strongly from the design principles outlined in this chapter. Such measurements are only relevant if tree belts have a sufficient width and, as detailed later, if they have developed a decomposing layer of leaf litter beneath them. Nevertheless, where possible, the parameters used in the numerical models are based on measured properties of vegetation, and care is taken to simulate settings that can be realised in practice, for example, with relation to biomass density. Furthermore, some major assumptions in the modelling process have been validated based on scale model experiments. When simplifications of the numerical procedures were needed to limit the computational cost, full-wave numerical techniques were used as a reference to safeguard their accuracy.

5.2.2 Acoustical effects operating in a vegetation belt shown to be additive

The acoustical effect of a vegetation belt is a joint effect of the vegetation-affected soil, a shrub under storey, stems, and foliage. An important research question was whether the acoustical effect of the different components could be considered and modelled separately given the complexity of the processes involved. The latter could not be answered a priori.

Based on numerical calculations, scale model studies, measurements, and discussions reported in literature, there is sufficient evidence that the different components in a vegetation belt contribute more or less independently to the total noise reduction achieved. This means that the acoustical effects of the different components in a tree/vegetation belt can be determined separately, and linearly added to find the total effect.

5.2.3 Interactions between sound waves and vegetation belts

5.2.3.1 Soil effect

The ground layer can lead to destructive interference between the direct contribution from source to receiver, and a ground-reflected sound path. This effect is often referred to as the acoustical *ground effect* or *ground dip*. The presence of vegetation leads to an acoustically very soft (porous) soil as a consequence of plant litter and plant rooting. This results in a shift in the ground effect towards lower frequencies compared to, for example, sound propagation over grassland.[2] Due to the typical source-receiver heights in road traffic noise applications, the attenuating ground effect mainly occurs below 1 kHz. In order to benefit from this ground effect, the specular reflection point should be located inside the vegetation belt. In traffic noise situations, with a vegetation belt close to the road, this is most often the case. Simple geometrical considerations using the concept of sound rays explain when soil effects can be exploited (see this volume, Chapter 6[3]).

5.2.3.2 Acoustic effects of shrubs and hedges

Shrubs (e.g., as under storey in a tree belt) will lead to multiple scattering, redirecting sound energy away from the (direct) sound path between source and receiver. In addition, it can lead to enhanced attenuation due to longer propagation paths and to an increased interaction with the soil. The effect of a shrub belt is expected to be rather limited because foliage scattering only affects high frequencies, whereas most of the acoustical energy from road traffic is contained in the low-frequency range. The insertion loss will depend on the vehicle speed, which influences the relative proportions of low and high frequencies. At higher vehicle speeds, more of the energy is emitted at high frequencies and foliage scattering becomes more relevant than at lower vehicle speeds. Denser and deeper shrub layers will produce a stronger positive effect. In combination with a tree belt, low-growing vegetation in theory provides additional noise reduction. However, shading of sunlight due to the tree canopies, and competition with trees for nutrients and water make such additional effects difficult to attain in practice.

A hedge is a row of closely spaced shrubs having a limited depth. Noise reduction by hedges has been measured in a number of field experiments. Different pass-by experiments along hedges with cars driving at speeds below 50 km/h were performed. A wide variety of hedges were selected, with depths between 1.8 and 2.5 m. Road traffic noise shielding was found to be from 1 dB(A) up to a maximum of 3 dB(A).[4] The higher noise reductions were observed in cases where one could also benefit from the soil effect close to or underneath the hedge. Hedges should have sufficient depth and should be as dense as possible. In addition, there should be a sufficient amount of biomass close to the ground. This is needed to prevent sound propagating underneath the hedge. This is especially important for rolling noise, which is generated very close to the road surface. The physical noise reduction provided by shrubs and hedges in road traffic noise applications, therefore, could be considered as limited in most cases.

5.2.3.3 Trunks

A belt of tree trunks can be modelled as a grid of cylinders, with their length axes orthogonal to the ground plane, leading to reflection, diffraction, and scattering of sound propagating through it. The absorption coefficient of bark covering the tree trunks is typically small. Due to the many interactions between sound waves and trunks' surfaces, however, it becomes a relevant parameter. The absorbing properties of tree bark have been measured.[5] There are some clear differences in bark absorption between species. A rather conservative value for the absorption coefficient ($\alpha = 0.075$) will be used in the calculations presented in this chapter, to prevent exaggerating this effect. The bark absorption coefficient seems to be rather independent of sound frequency and therefore can be represented by a single value. Irregularities at the bark surface could further increase absorption and scattering, especially at higher sound frequencies.

Trunks, and also the soil effect, interact with sound frequencies below 1 kHz. In addition, the multiple scattering process in-between the trunks can lead to specific filtering effects, namely, the presence of stop and pass bands. Such processes are governed by the combination of trunk diameter and trunk spacing. In some frequency bands ("stop bands"), there is a strong reduction of the noise level. In other bands ("pass bands"), the sound waves are hardly affected and, therefore, strongly transmitted through the trunk layer. These effects occur at very specific sound frequencies. As a result, the noise reduction spectrum has a peaky and rather complex character. Stop and pass bands caused by trees have been observed as reported by Martinez-Sala et al.[6] However, such band gap effects are expected to be of minor importance for road traffic noise reduction as the trunk-filling fractions where such effects appear exceed those found in tree belts.

A trunk will reduce higher sound frequencies by direct shielding. In general, shielding becomes important when the wave lengths are small relative to the dimensions of an object in the propagation path.

5.2.3.4 Crown scattering

When both the source and receiver are located below the canopy level, scattering by crowns results in a slightly negative effect. Sound energy emitted upward can be scattered downward by the canopy toward a receiver located behind the crown layer. In absence of tree canopies, such sound paths would continue upwardly without reaching receivers at ground level. Note that scattering produces the opposite effect when source and receiver are at the same height as the foliage.

The crown scattering effect will be most prominent in situations with high vehicle speeds that are dominated by higher frequencies. Measurements behind a noise wall along a highway showed that, at a sound frequency of 10 kHz, downward scattering from a single row of tall trees adds 6 dB.[7] Below 2 kHz, effects are smaller than 1 dB. As a result, the total effect for A-weighted road traffic noise is expected to be much lower than 1 dB(A) as well, making crown scattering a minor (but negative) effect.

5.2.4 Planting schemes for tree belts

5.2.4.1 Basic behaviour

The fraction of the ground area (as seen in plan view) taken by the tree trunks' cross sections (measured at breast height), hereafter referred to as the *trunk basal area*, is the basic parameter influencing the acoustical shielding of a tree belt. With increasing basal area, the acoustical shielding increases. High basal areas can be obtained with large stem diameters and by limiting the distance between adjacent trees. When analysing the scatter plot (Figure 5.1) relating road traffic noise insertion loss and basal area, there are clearly important modifiers. At all trunk basal areas, the planting scheme and specific trunk spacing/diameter combination have an important influence on the noise shielding. In this section, interesting choices and combinations will be identified.

Clearly, there are some practical limitations since one works with living material. A basal area of 1% is considered to be rather easily achievable with common species and maintenance.[8] However, if higher basal areas are needed, selecting for specific species as *populus* or *salix* might be needed. In addition, specific maintenance like pollarding (cyclical removing of all branches) contributes to increased stem diameters even at close spacing. In all cases, calculations have been limited to a basal area of 2%.

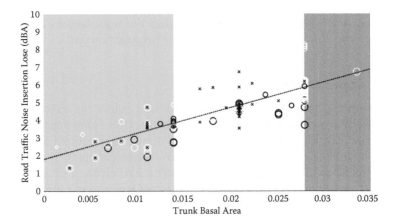

Figure 5.1 Scatter plot between basal area and road traffic noise insertion loss of a 15-m-deep tree belt. Each marker stands for a particular combination of planting scheme and trunk diameter. Three regions have been identified with relation to the trunk basal area. Green indicates that the tree belt can be easily achieved with common species. Yellow means that specific species must be selected and specific maintenance is needed. Red indicates that the tree belt probably will be difficult to realize with living material. The best fitted linear curve through all numerical results is drawn. The insertion loss is defined relative to sound propagation over grassland as illustrated in Figure 5.2. Following traffic parameters have been used. A four-lane road is considered, with a fixed vehicle speed of 70 km/h. The traffic is composed of 85% light vehicles and 15% heavy vehicles. A receiver height of 1.5 m is considered behind the belt. The diameter of the circles is proportional to the diameter of the trunks modelled. The black circles denote regularly spaced schemes; the white circles indicate that some randomness is involved. The crosses indicate predictions where part of the trees were omitted.

The numerical calculations shown in Figure 5.2 start from a practical baseline situation, namely, grassland along a four-lane road to be partly replaced by a non-deep tree belt. A receiver is located behind the tree belt. Clearly, in the reference situation (sound propagation from the road over grassland toward the receiver), there are already acoustical effects operating, such as a change in soil characteristics at the interface between the road and grassland, and the soil effect one obtains from grassland.

5.2.4.1.1 Tree spacing

Tree spacing has a rather straightforward effect. When decreasing the distance in-between neighbouring trees, the basal area increases and so does the noise shielding. For a tree spacing larger than 3 m, the main effect is expected to be the soil effect, unless the trunks have very large stem diameters or when the spacing along the road length axis is kept limited (see Section 5.2.4.2.1, Rectangular Schemes Are Preferred).

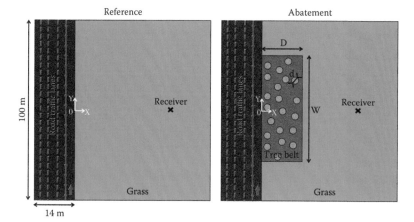

Figure 5.2 Reference case and abatement case considered for the numerical predictions reported in Figure 5.1. When modelling a single car pass-by, as shown in Figure 5.4, only the right traffic lane has been used. D indicates the depth of the tree belt (orthogonal to the road length axis), W the width (along the road length axis), and d is the diameter of the individual trees.

5.2.4.1.2 Trunk diameter

Noise shielding is proportional to the average trunk diameter in case of a given spacing. This relation is of a linear-quadratic nature. At high trunk diameters, there will be a rapid increase in the acoustical efficiency when further increasing the trunk diameter. At small trunk diameters, the efficiency is more or less linearly dependent on the diameter.

The curve presented in Figure 5.3(c) also can be considered as the evolution over time, starting from a tree belt initially consisting of small diameters. It is then assumed that the diameters of all trees grow at an equal rate. The maximum achievable diameter will depend on the tree species and on specific stand characteristics. Here, the noise reduction is shown when assuming that such diameters are possible. Another assumption is that there is no time of evolution in the characteristics of the soft forest floor; it is assumed that directly after planting, one can benefit from its full effect.

5.2.4.1.3 Depth of the tree belt normal to the road

With increasing depth of the belt, perpendicular to the road length axis, noise is reduced further. However, already for non-deep (but optimised) tree belts, in combination with the presence of a soft forest floor, substantial noise reduction is possible. In case of a fixed receiver position, the noise reduction increases linearly with the depth of the tree belt, when assuming

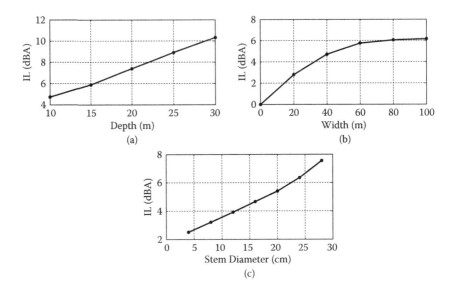

Figure 5.3 Road traffic noise insertion loss (IL) in function of depth (a), width (b), and stem diameter (c) of a tree belt. The reference is sound propagation over grassland. The planting scheme used is a rectangular one (1 m spacing along the road length axis, 2 m normal to it). Random displacements from the regular grid were allowed at 25% of the spacing in both directions. Both light and heavy vehicles drive at 70 km/h, the heavy vehicle fraction in the traffic composition is 15%. In (a) and (b), a tree diameter of 22 cm is chosen. In (a) and (c), the width of the tree belt was 100 m. In (b) and (c), the depth of the tree belt was 15 m. Receivers are located at 40 m from the border of the road in (a), at 30 m in (b), and at 50 m in (c). The receiver height was 1.5 m in all cases.

the belt is sufficiently wide to prevent sound encroaching from either sides. Such non-deep tree belts have a wider applicability in urban and suburban environments because they take up less land, and also can be used, for example, to border an urban park along a road.

5.2.4.1.4 Trunk height

Trunk height was shown to be a rather unimportant parameter, given the fact that road traffic sources are typically at very low heights above the road surface, and as long as the belt is located close to the road. Only at very low trunk heights, an increase of the acoustical effect with height has been observed in numerical calculations (when neglecting crown scattering). When the source and receiver are located below the bottom of the canopy, slightly negative effects might appear by downward scattering. When both the source and receiver are near the canopy height, positive effects are expected, however.

5.2.4.1.5 Width of the tree belt

With decreasing width of the tree belt (along the road length axis), the global efficiency for road traffic noise shielding decreases. In case of belts of limited width, noise from a passing car will only be reduced during the time period when cars are behind the tree belt (from the point of view of the receiver). When the cars become visible for the receiver (which means oblique sound paths that do not interact with the tree belt), a similar situation as in absence of the tree belt is obtained. The global effect of the tree belt is then the combination of the contributions of all possible source positions along the road. Due to the nature of the decibel scale, a strong reduction at some road segments combined with a limited reduction at others will lead to a rather limited global noise abatement efficiency (relative to the reference situation).

Note, however, that oblique sound paths, from sources farther away from the receiver, arrive at the receiver with less energy due to geometrical spreading, atmospheric absorption, and a longer interaction zone with grassland. Clearly, at some point, making the tree belt wider is not efficient anymore. The optimal width of the belt also depends on the distance between the belt and the receiver. Receivers at close distance need much smaller belt widths compared to receivers at greater distances.

Similar to guidelines as concerns the needed length of classical noise walls, the angle enclosed between the receiver and both edges of the tree belt should exceed 140 degrees. Numerical analysis showed that when making the tree belt wider, road traffic noise shielding hardly increases.

The influence of the width of the tree belt is illustrated by means of a simulated pass-by of a single car on the closest lane in Figure 5.4. The depth of the belt is taken to be 15 m, and the position of the receiver is fixed (at 30 m from the border of the road). Road segments between y = −50 m and y = 50 m are considered. Tree belt widths ranging from 20 m to 100 m (i.e., fully covering the road section under study) are considered. When the car is visible from the viewpoint of the receiver, noise abatement at these positions is very limited, and the level curve is very close to the one of the reference case. When the source is not visible, the sound pressure levels drop drastically. For the widest tree belts, all sound paths interact with the vegetation belt, resulting in a noise reduction at all source positions. When tree belts are wide, the source locations near the beginning and end of the tree belt have the longest interaction path and the improvement in noise reduction relative to grassland then is largest.

5.2.4.1.6 Distance of receiver relative to the tree belt

If a tree belt is sufficiently long, increasing the distance between the receiver and the road (perpendicular to it) will not reduce the noise shielding efficiency. The reason for this is that the noise abatement is obtained during transmission through the belt. This is clearly different from the attenuation

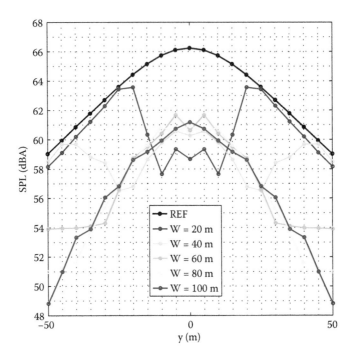

Figure 5.4 Sound pressure level (SPL) contribution from the different source positions along the road. REF indicates the reference situation, i.e., sound propagation over grassland. Different widths of tree belts are simulated. A single light vehicle, driving at a constant speed of 70 km/h, is assumed on the closest lane (relative to the receiver, see Figure 5.2). A pseudo-random spacing of 1 m along the road axis, and 2 m orthogonal to it, is simulated. The random displacement from the regular grid was 25% of the spacing in both directions. A tree diameter of 22 cm was chosen and a receiver height of 1.5 m.

mechanism of a traditional noise wall, where diffraction over the top edge becomes the main sound propagation path. The larger the detour sound has to make relative to a virtual direct sound path between source and receiver, the more efficient a noise wall will be. At greater distances from the wall, however, this detour will be limited and the shielding rendered ineffective. In contrast, noise abatement from a tree belt is maintained at larger distance. Note, however, that with increasing distance relative to the road, absolute sound pressure levels decrease due to geometrical spreading of sound, atmospheric absorption, and interaction with grassland both in the case of a tree belt and noise wall. The need for a noise abatement solution, therefore, could be less pronounced at larger distances. In case of a finite-width belt parallel to the road, contributions from sound paths not interacting with the belt could result in a decreased efficiency when increasing receiver distance, as explained before. The greater the distance between the receiver and the belt, the stronger the importance of such sound paths. So greater receiver distances might need wider tree belts. The same argument applies to the length of a noise wall.

5.2.4.2 Toward higher noise abatements

The use of specific planting schemes can have a strong influence on the shielding, as illustrated by the scatter plot presented in Figure 5.1. Even in case of a fixed trunk basal area, there can be quite some variation in shielding.

In this section, different planting schemes are studied. Of special interest is the choice of a specific lattice (while keeping the number of trees per unit area constant), the introduction of (some) randomness in the tree belt, and the presence of gaps. Such details in the planting scheme can enhance noise reduction. Such choices become increasingly important when the tree density increases. For very sparse tree belts, such additional positive effects will not be observed.

5.2.4.2.1 Rectangular schemes are preferred

The choice of a specific lattice, while keeping the number of trees per unit area constant, can result in additional shielding. Various planting schemes have been numerically evaluated, and rectangular schemes were shown to be most effective given that the smallest spacing is along the road length axis. This will improve the noise shielding compared to the same scheme, but rotated over 90 degrees. Increasing the spacing perpendicular to the road, while keeping the spacing parallel to the road fixed and small, lowers the noise shielding only to a limited extent. With increasing tree belt density, the influence of orientation in rectangular schemes becomes enhanced. In a rectangular scheme with a spacing of 1 m × 2 m, a rotation of the planting scheme over 90 degrees can result in a difference in shielding exceeding 1.5 dB(A).

It can be concluded that dense rows of tree trunks parallel to the road length axis are interesting. At the same time, the spacing in-between such rows (normal to the road length axis) can be increased without significantly decreasing the noise shielding. This has the advantage that the average tree density becomes lower and such tree belts become easier to realize. Such wider paths parallel to the road facilitates the use of vehicles for planting, inspecting, and maintaining trees. Compaction of the soil, however, should be avoided to maintain the forest floor effect.

5.2.4.2.2 Benefits of introducing some randomness

Small deviations from a perfectly ordered positioning of trees, following a particular grid, could lead to an increase in noise shielding.

Numerical simulations show that an optimal effect is found at a randomness near 20% of the spacing in case of a 1 m × 2 m rectangular scheme with uniform stem diameters of 22 cm. This means that each tree is moved from its regular position and placed randomly in the zone equal to one fifth of the distance in-between the trees. These shifts are introduced both along and perpendicular to the road length axis. Already small random displacements result in an increase in shielding relative to fully regular schemes. The curve relating the noise shielding and the degree of random displacement shows a rather broad maximum near 10–25% (Figure 5.5). Fully random placement, while using the same number of trees per unit area, gives more shielding than fully ordered schemes with the same trunk basal area. There is, however, a slight preference for small disturbances. A possible reason for the positive effect of the small displacements is that periodicity effects, although not very pronounced, are still present, while (negative) focussing effects disappear.

Clearly, since randomness is involved, different realisations do not yield exactly the same tree grids and lead to some variation in the noise shielding. This variation is illustrated by the error bars having a total length of two times the standard deviation. Differences up to 0.5 dB(A) are possible between the different realisations for a given degree of randomness.

Note that when the basal area could be increased (either by natural or artificial means), the effect of including pseudorandomness, relative to fully ordered schemes, is strongly enhanced.

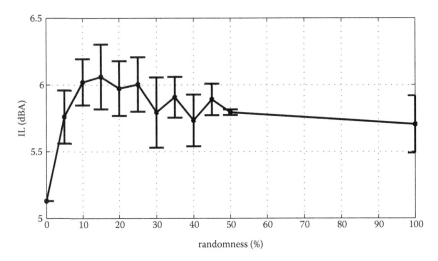

Figure 5.5 Road traffic noise insertion loss in function of increasingly allowing randomness as for the stem centre location. The schemes used are equivalent to a 1 m × 2 m scheme, with stem diameter of 22 cm. Both light and heavy vehicles drive at 70 km/h; the heavy vehicle fraction in the traffic composition is 15%. The receiver is located at 40 m from the border of the four-lane road, behind a 15-m-deep tree belt, at a receiver height of 1.5 m.

Avoiding a uniform trunk diameter is another way to increase noise shielding. The effect of slightly disordered trunk centre locations is predicted to be somewhat stronger than variations in trunk diameter relative to fully uniform tree grids. No additional positive effects of both randomness in stem positioning and diameter are predicted.

Allowing a small degree of randomness in both stem centre location and stem diameter is interesting from a practical point of view, and is likely to occur by nature. However, keeping control of the deviations could be relevant in the viewpoint of further optimising noise shielding.

5.2.4.2.3 Benefits from gaps

Omitting a number of trees from a structured tree belt does not necessarily mean that the noise shielding will significantly decrease. However, care is needed when creating such gaps.

Omitting full rows of trees, parallel to the road, hardly affects the acoustical shielding. It is assumed that near these left-out rows in the belts, the soft forest floor still develops. The first number of rows at both the source and receiver side should be left unaltered. This is illustrated by the values in Table 5.1 and the corresponding Figure 5.6, showing what rows were left out.

Omitting full rows of trees, normal to the road, will negatively affect the noise shielding by the belt because this action might produce channels for the noise to propagate almost undisturbedly from specific source positions on the road toward receivers. Depending on the location of these channels, the noise reduction can be strongly reduced.

Randomly omitting trees within the belt slightly lowers the noise shielding. Consequently, thinning of not well-developing trees after planting should not be a problem from the viewpoint of noise reduction; of course, within reasonable limits. The effect on road traffic noise insertion loss of increasingly removing trees, in a random way, from an originally regular belt is illustrated in Figure 5.7. In this configuration, removing up to 25% of the trees hardly affects the noise reduction.

Clearly, these findings are interesting for the practical realisation of tree stands, knowing that dense stands should be strived for to yield sufficiently large effects. The presence of gaps lowers the average tree density and ensures that more resources (light, water, nutrients) are available for growth. Therefore, trees planted in densely positioned rows, followed by open spaces, is a practical solution.

A summary of useful actions to increase noise shielding by vegetation belts is shown in Figure 5.8. In Figure 5.9, some guidelines are provided to reduce the tree density without significantly lowering the acoustical shielding. In the latter, there are clearly constraints as regards the amount of trees to be omitted as discussed before.

Table 5.1 Calculated road traffic noise insertion losses for 15-m-deep tree belts using rectangular or square planting schemes (= $IL_{regular}$)

Spacing a (m)	Spacing b (m)	Trunk diameter (cm)	ILregular (dB(A))	ILrandom (dB(A))	Basal area
1	2	22	5.2	6.0	0.020
2	1	22	3.7	5.6	0.020
2	2	22	2.8	3.3	0.010
2	3	22	1.9	2.7	0.008
3	2	22	2.9	3.1	0.007
3	1	22	3.9	3.9	0.013
1	3	22	4.4	4.8	0.015
3	3	22	2.4	2.5	0.005
1	1	11	3.8	4.8	0.010
0.5	1	11	5.9	8.1	0.020
0.5	1	11		6.7[*]	0.015
0.75	0.75	11		5.8[*]	0.013
1	2	22	4.5[a]	5.8[a]	0.015
1	2	22	4.5[b]		0.015
1	2	22	4.9[c]		0.015

Note: Spacing *a* is the distance in between the centres of the trunks along the road length axis; spacing *b* is the spacing normal to the road length axis. Pseudorandomness indicates random shifts in trunk location of at maximum 25% of the spacing (= IL_{random}). The following traffic parameters have been used: 15% heavy vehicles and all vehicles, equally distributed over the four lanes, drive at 70 km/h. The receiver height is 1.5 m in all cases.

[*]   25% of the trees have been omitted.
[a-c] See Figure 5.6.

5.3 IMPROVING MICROCLIMATOLOGY BY VEGETATION

Sound propagation outdoors can be strongly influenced by meteorological conditions, especially at larger distances from the source. At 600 m from the source, the variation in noise level due to changes in meteorology could be as large as 40 dB(A).[9] This difference could mean that a particular source is either heard distinctly or not at all. Close to the source, meteorologically induced variations in the received sound pressure level are much smaller.

Of special interest is refraction of sound caused by changes in air temperature and wind speed with height in the atmospheric boundary layer. In the absence of refraction, sound propagation can be represented by straight sound rays. In contrast, if there is refraction, sound rays will be bent. This bending can be either upwards or downwards. Downward bending leads to increased sound pressure levels at ground level, thus, in the zone where receivers are expected. Essentially, a larger portion of the sound

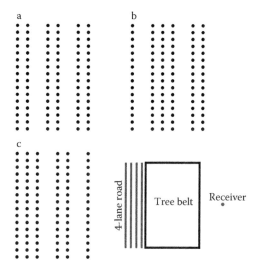

Figure 5.6 Some ways considered for removing two rows (parallel to the road length axis) out of eight.

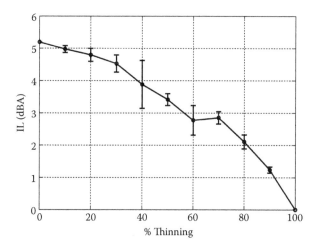

Figure 5.7 Road traffic noise insertion loss in function of increasingly thinning the tree belt (in a random way). The scheme at 0% is a fully populated rectangular one, with a spacing of 1 m along the road length axis and 2 m normal to it. The stem diameter is 22 cm. Both light and heavy vehicles drive at 70 km/h; the heavy vehicle fraction in the traffic composition is 15%. The receiver is located at 40 m from the border of the four-lane road, behind a 15-m-deep tree belt, at a receiver height of 1.5 m. The error bars have a length of two times the standard deviation and are the results of various (random) realisations of a given thinning percentage.

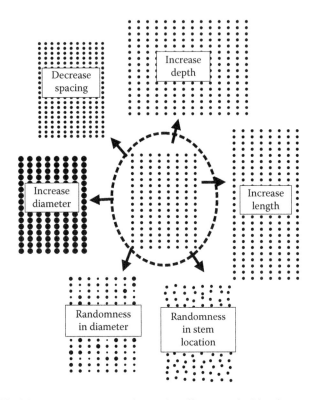

Figure 5.8 Useful actions to increase the road traffic noise shielding by a tree belt.

energy contained in the source reaches receivers instead of being equally distributed in all directions (as occurs, for example, for point sources in a non-refracting atmosphere). Upward bending will lead to lower sound pressure levels at receivers at the ground level. The effects of refraction are of special interest in typical road traffic noise situations, where both sources and receivers are close to the ground's surface.

A ground-based temperature inversion, typically occurring at night under clear sky conditions following strong solar radiation during the daytime, is a case where downward bending is observed, leading to increased sound pressure levels. Downwind sound propagation results in a similar effect. Upwind, strongly decreased sound pressure levels will be observed relative to a windless situation. A similar effect occurs during daytime in case of strong solar radiation, leading to a decrease of temperature with height and, consequently, upward refraction.

Wind effects can be very pronounced near noise barriers. Especially in case of vertically erected noise walls, a strong decrease in shielding for downwind receivers is observed.[10–12] Similar effects appear in acoustic shadow zones

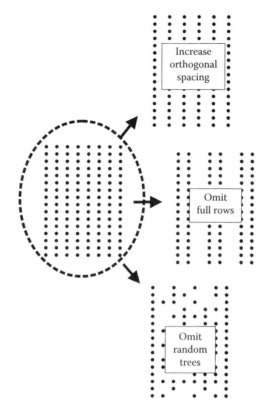

Figure 5.9 Useful actions to decrease the average trunk basal area of a tree belt without significantly lowering the road traffic noise shielding.

formed behind a non-aerodynamically shaped object (e.g., near a building). This effect occurs at close distance behind a barrier in the zone where we would expect high shielding in absence of wind. With increasing wind speed, refraction effects can be quite dramatic, easily halving the barrier efficiency in road traffic noise situations. This is one of the reasons why noise walls often fail to live up to the expectations of people living nearby.

Vegetation influences microclimatology and, therefore, could be helpful in this respect. The shading properties of vegetation mean that temperature profiles will be altered. Also, the canopies of trees or shrubs can be efficient windbreaks. This idea has been used over the centuries in agricultural practice to give shelter to animals or to improve crop productivity. By changing the microclimatology, the sound propagation conditions can be changed in turn. It will be shown in the following sections in what type of cases, and to what extent, vegetation can be used to achieve sound propagation conditions leading to decreased levels. Of special interest are limiting nocturnal ground-based temperature inversion conditions with strips of forests, and reducing downwind refraction near noise barriers with rows of trees.

5.3.1 Reducing nocturnal temperature inversion effects

After sunset, the soil typically cools down much faster than the layers of air in the atmospheric boundary layer. This leads to an increase of air temperature with height, relative to the temperature at ground level. This results in downward bending of sound and, thus, larger sound pressure levels compared to sound propagation in a homogenous atmosphere (meaning a constant air temperature with height in the lowest part of the atmosphere). The presence of vegetation will prevent the development of such a ground-based nightly temperature inversion layer. When vegetated, the soil is not able to freely radiate to the sky, and part of the heat is trapped in the zone below the canopy.

Numerical predictions show that a 50-m-deep forest strip may strongly decrease night noise levels in situations with a nocturnal inversion layer. At a receiver height of 2 m, and at a distance of 200 m from the road edge, this benefit could easily exceed 10 dB(A) compared to sound propagation over grass-covered soil. The state of the atmosphere, which is changed in a positive way from the viewpoint of noise abatement, comes on top of the noise shielding provided by the combination of the soft forest floor and the presence of trunks. With increasing receiver height, the influence of refraction on the sound pressure levels, in general, is smaller, and the gain by the presence of a strip of forest, consequently, will be smaller as well.

During daytime, on the other hand, the presence of a tree belt results in a slightly worse atmospheric condition. Due to the shading provided by the canopies, the zone close to the soil heats up much slower than the layers of air above the canopy. In this way, a temperature inversion condition is created. On the other hand, in an open non-vegetated environment, an upwardly refracting atmosphere is built up as bare soil typically heats up much faster than the air. This means that the positive action of the upward bending of sound is (partly) counteracted by the presence of vegetation.

Note, however, that in case of a strong temperature decrease with height, a shadow zone will be formed at some distance from the source, depending on the height of the source and receiver. Sound energy can reach this zone by turbulent scattering only. As a result, sound pressure levels will be very limited in this zone. This means that, in practice, the slight increase in sound pressure levels due to the presence of the vegetation may be of limited importance. Moreover, the presence of the shielding by the forest floor and aboveground biomass will counteract this negative refraction effect. In total, the presence of a strip of forest could lead to an increase of a few dB(A) relative to the sound level at the same distance over open grassland.

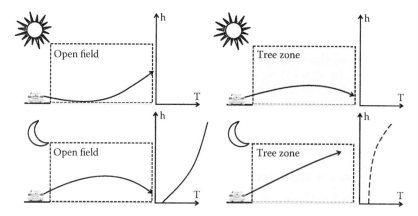

Figure 5.10 Illustration of how sound propagation is influenced by vertical air temperature profiles in an open field and in a forested zone, during day and night. Idealised air temperature profiles are shown under clear sky conditions and in the absence of wind. It is schematically depicted how sound rays could be bent in such cases.

Similarly to the change in micrometeorology at night, this effect will be most pronounced for receivers at limited height.

It can be concluded that strips of forest near roads, especially when they become deeper, could significantly change the state of the atmosphere, influencing, in turn, the sound propagation conditions. As argued above, the positive influence during the night is expected to outperform the negative action during the day. It has to be noted that such temperature-related refraction effects will be most pronounced under clear sky conditions both at night and during the day. The effects described in this section are illustrated in Figure 5.10.

5.3.2 Reducing wind effect near noise barriers

5.3.2.1 Real-life proof of concept

A real-life field experiment along a highway, described in detail by Van Renterghem and Botteldooren,[7] has shown a significant positive effect (at close distance) of a row of trees behind noise barriers. A 4-m-high noise walls on either side of the highway was present, with tall trees behind the noise walls at the receiver side. The tree canopies extended about two times the barrier height above the noise wall. The canopy started somewhat below the noise barrier top. Simultaneous and continuous measurements were performed behind a part of the noise barrier with trees and another part close by without trees. Onsite wind measurements were made. Such a setup allowed for a direct comparison of the tree effect in downwind conditions. With increasing wind speed, the positive effect of the presence of trees increased linearly. At a wind speed of 12 m/s at a height of 10 m, almost 4 dB(A) of the noise shielding due to the barrier was recovered by the presence of the trees. Furthermore, this effect was shown not to be restricted to the exact downwind direction. When averaging over wind directions up to 45 degrees away from the exact downwind direction, only 0.5 dB(A) was lost of this positive effect. In case of upwind sound propagation, i.e., when the wind was blowing from the receiver toward the road, sound levels were only slightly affected by the presence of the trees.

5.3.2.2 Validated numerical model

The effect of many configurations of windbreaks, near single noise walls and noise walls at either side of a line source, has been measured at scale in a wind tunnel.[13] It was found that the specific action of the presence of the windbreaks, which will be referred to as the *tree effect*, was well predicted by using a combination of flow modelling computational fluid dynamics (CFD) and a full-wave sound propagation model taking into account in detail the influence of the flow finite-difference time-domain technique (FDTD).[14] In this chapter, some guidelines are provided for the use of vegetation near noise barriers to reduce wind effects. These are based on predictions from this numerical model.

5.3.2.3 Worst-case scenario

Only downwind sound propagation is considered, which corresponds to a worst-case scenario. Although the exact downwind conditions (wind direction normally incident, relative to the barrier) might only occur a fraction of the time, the negative effects also are present in case of obliquely propagating sound paths in realistic road traffic noise situations.[7] Furthermore, these downwind conditions could become dominant when calculating equivalent sound pressure levels over a longer period.

5.3.2.4 Quantifying the windbreak effect of canopies

Predicting the windbreak effect of trees is complex, and has been simplified to a single parameter, namely, the pressure resistance coefficient k_r. The larger this value, the stronger is the pressure drop over the canopy and the more the wind speed is reduced. The pressure resistance coefficient can be linked to quantifiable and documented tree parameters,[15] and is the product of the drag coefficient C_d of the plant elements, the leaf area density (LAD), and the thickness of the canopy (along the wind direction) D_x:

$$k_r = C_d \cdot LAD \cdot D_x \tag{5.1}$$

The drag coefficient is a general aerodynamic parameter. The larger its value, the less aerodynamic and, thus, potentially more reducing wind speed. A drop-like form has a very low value (<0.1), while a cube has a much larger value (near 1). Deciduous trees have smaller drag coefficients (0.2) than coniferous trees. However, there are also what is called "windbreak" deciduous species[16] (e.g., tamarisk, elm, and American plum), approaching the behaviour of coniferous trees (e.g., spruces, pines, and junipers with drag coefficients close to 1).

The LAD, or the equivalent needle area density (NAD) for conifers, is another important parameter in quantitative plant research. This parameter usually depends largely on the height within the crown. Typical values of the maximum LAD of the canopy of deciduous trees is $1 \text{ m}^2\text{m}^{-3}$. The NAD is usually larger, and ranges up to $2 \text{ m}^2\text{m}^{-3}$. However, there is a lot of variability depending, for example, on species, stand characteristics, and age.

5.3.2.5 Optimal designs

Based on many calculations with detailed flow and acoustic models, it was concluded that moderate to dense canopies are needed to maintain the noise shielding under downwind conditions. The negative action of the wind becomes more pronounced at higher frequencies and, consequently, near roads with higher vehicle speeds. This also means that the positive action of trees can be greater in such cases.

5.3.2.5.1 Single noise walls: leaving gaps or not?

The use of trees to reduce wind effects seems especially successful near single noise walls. The distinction has to be made whether the canopy starts below or at the top of the noise screen, or whether a gap in-between the bottom of the canopy and the top of the screen is present. This choice will determine where the largest reduction of refraction will occur in downwind direction.

The absence of gaps improves shielding at close distance behind the barrier in wind. With increasing downwind distance, effects become smaller and the canopies might exhibit a negative effect.[17] The reason for the latter is that the driving force for the wind effect near barriers, namely, the zone with strong gradients in the wind speed, is moved upwards. Sources at larger distances from the barrier (upwind), or receivers at larger distances downwind, interact with such zones, and this could lead to a negative effect (compared to the case of a barrier without trees in wind). Note, however, that sound interacting with these higher zones travels a longer distance, yielding lower levels anyway by other propagation effects, such as geometrical spreading and atmospheric absorption. At these distances, barrier efficiency is limited anyhow and other noise sources could potentially become important. Another consequence is that when the source is located close to the barrier, the positive action of the trees is larger and will be present up to a longer distance. When the sound source is farther away from the barrier, trees will be less effective in reducing refraction effects.

The presence of gaps in-between the barrier and the bottom of the canopy is another option. This means that, at a close distance behind the barrier, the wind effect is not recovered and the strong decrease in shielding by the action of the wind remains.[17] However, the positive action of the trees increases with distance, and will be observed up to a greater distance downwind.

This is illustrated by the simulation results in Table 5.2. A thin noise screen with a height of 4 m is placed near a four-lane road. A row of trees with a top height of 12 m is located directly downwind from the barrier. Receivers are located in a zone up to 250 m downwind, and, from ground level, up to a height of 4 m. The presence of a gap (of 1 m) between barrier and canopy bottom leads to an overall best situation. For a moderate incident wind speed, a significant amount of the shielding that was lost by the wind has been recovered by the presence of a row of trees (for light vehicles driving at 110 km/h). At a vehicle speed of 70 km/h, effects become somewhat more moderate. In case of strong winds, this recovery percentage becomes smaller. Leaving no gaps between the canopy and the barrier top yields limited recovery when averaged over all distances considered. However, if the zone of interest is close to the barrier, gaps become less efficient.

With increasing top height of the canopy, more wind reduction and more recovery of the shielding is expected. The influence of canopy height becomes more pronounced with increasing vehicle speed. Simulations comparing top heights of 6 m, 8 m, and 12 m (no gaps) near a 4-m-high noise screen show that, in the case of moderately dense canopies, differences in effects are rather limited. In the case of dense canopies, additional height has a clear positive effect. The higher the top of the canopy, the larger is the upward shift of the zone leading to downward refraction. This is clearly positive for the shielding. In order to still interact with this zone, sources have to be located farther away, or only receivers at much greater distances downwind could be affected.

5.3.2.5.2 Rigid noise walls on either side of the traffic
lanes: moderate improvements

In case of fully reflecting noise screens on either side of the source, strong improvements by the presence of trees are expected at short downwind distances only. With increasing wind speed, this effect is even enhanced. On the other hand, negative effects related to the placement of trees are predicted at larger downwind distances. For moderate winds, only a small amount of the shielding that was lost by the wind can be recovered by placing trees (with canopies extending 8 m above the screen top) averaged over a zone up to 250 m. Dense canopies are preferred. In strong winds,

Table 5.2 Average predicted effects (in dB(A)) in different zones downwind from the (downwind) noise wall (expressed in screen heights H, which is 4 m in the current cases, and averaged over receiver heights between ground level and 4 m) in a highway configuration

	3H–13H	13H–23H	23H–33H	33H–43H	43H–53H
Single noise wall					
(vehicle speed = 70 km/h)					
IL (no wind)	13.6	9.9	8.3	7.5	7.0
SIROS (moderate wind)	2.5	3.9	4.6	4.7	4.7
TE (dense tall canopies, no gaps, moderate wind)	1.3	2.0	1.5	0.3	−0.8
TE (dense tall canopies, gap, moderate wind)	0.9	2.0	2.1	1.5	0.8
(vehicle speed = 110 km/h)					
IL (no wind)	15.9	12.1	10.0	8.8	8.0
SIROS (moderate wind)	3.4	5.3	6.3	6.6	6.6
TE (dense tall canopies, no gaps, moderate wind)	1.6	2.6	2.5	1.3	−0.2
TE (dense tall canopies, gap, moderate wind)	1.0	2.7	3.2	2.7	1.9
Rigid noise walls at both sides					
(vehicle speed = 70 km/h)					
IL (no wind)	7.1	2.9	0.9	−0.2	−0.8
SIROS (moderate wind)	3.2	5.8	6.8	6.9	6.7
TE (dense tall canopies near both noise walls, no gaps, moderate wind)	2.2	3.1	1.8	−0.4	−2.4
(vehicle speed = 110 km/h)					
IL (no wind)	7.8	3.1	0.4	−1.3	−2.3

Continued

Table 5.2 (Continued) Average predicted effects (in dB(A)) in different zones downwind from the (downwind) noise wall (expressed in screen heights H, which is 4 m in the current cases, and averaged over receiver heights between ground level and 4 m) in a highway configuration

	3H–13H	13H–23H	23H–33H	33H–43H	43H–53H
SIROS (moderate wind)	3.8	6.8	7.7	7.6	7.1
TE (dense tall canopies near both noise walls, no gaps, moderate wind)	2.7	3.7	2.0	−0.6	−2.8

Note: IL is the noise reduction provided by the noise wall in absence of wind. A positive value means noise reduction relative to unscreened grass-covered ground. SIROS is the magnitude of the negative action caused by the wind. A positive value means that, during downwind sound propagation, a decreased shielding is predicted relative to the situation with the noise wall in absence of wind. TE is the tree effect, or the decibels that can be recovered due to the windbreaking effect of the canopies. A positive value means that trees are effective in reducing downward refraction at a given distance. Grassland is modelled downwind a four-lane highway. A moderate wind is assumed in all cases (represented by an incident wind speed of 6.9 m/s at a height of 10 m). The noise walls are modelled as rigid.

the global performance of placing trees near such perfectly reflecting screens is predicted to be negative. The choice whether to place trees in the case of rigid noise walls will depend on the distance of the dwellings behind the barrier and local wind characteristics. However, good practice demands absorption at the surfaces of parallel noise walls. Consequently, similar effects, as observed near single noise screens, are expected in cases of absorbing noise walls on either side of the traffic lanes.

5.3.2.5.3 Increased downward scattering of sound in the absence of wind

A possible drawback of placing a row of trees behind a noise barrier is the increased scattering of sound for receivers below the bottom of the canopy. However, scattering of sound is typically a high-frequency phenomenon, affecting part of the road traffic noise spectrum containing less energy. Measurements showed that at 10 kHz, the sound pressure level could be 6 dB higher because of tall trees behind a noise wall (compared to the same noise wall in absence of trees).[7] The effect on total A-weighted levels for highway noise has been estimated to be between 0.5 and 1 dB(A). Therefore, the positive action of trees in downwind periods is expected to be more important than the negative effect of downward scattering in a long-term assessment. Due to the nature of the decibel scale and human hearing, the increased noise levels during periods with downwind propagation could become dominant for equivalent sound pressure levels over a substantial period.

5.3.2.5.4 Steep berms: no trees needed

Nonsteep berms strongly reduce the wind effects compared to vertically erected noise walls due to their aerodynamic shape.[18] Steep berms, on the other hand, can still lose a lot of their noise shielding under downwind conditions, although somewhat less pronounced than in case of a vertically erected noise wall with a similar height. In contrast to their positive effects near noise walls, reducing wind effects by placing trees near steep berms is predicted to be rather unsuccessful. If trees are to be placed anyway, it is better to locate them on the source side, where they could partly limit negative effects of crown scattering toward receivers positioned below the canopy height. Given the limited improvement during windy periods, scattering could be relevant when looking at the global effect. However, vegetation on a berm ensures that its surface becomes acoustically soft, which is clearly positive for sound waves diffracting over the berm[18].

The optimal canopy designs to reduce downwind effects near a single noise wall, a steep berm, and noise walls on either side of a highway are summarized in Figure 5.11.

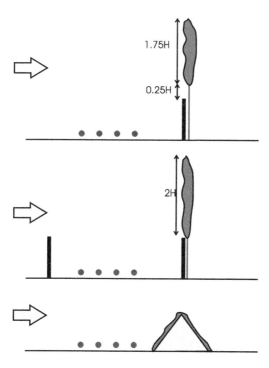

Figure 5.11 Optimal canopy design to reduce downwind effects near a single noise wall, a steep berm, and noise walls positioned at both sides of a four-lane highway. H indicates the noise wall height.

REFERENCES

1. Defrance, J., P. Jean, F. Koussa, T. Van Renterghem, J. Kang, and J. Smyrnowa. 2014. Innovative barriers. In *Environmental methods for transport noise reduction*, eds. M. E. Nilsson, R. Klæboe, and J. Bengtsson (pp. 19–47). Oxford, U.K.: Spon Press.
2. Huisman, W. H. T., and K. Attenborough. 1991. Reverberation and attenuation in a pine forest. *Journal of the Acoustical Society of America* 90(5): 2664–2677.
3. Attenborough, K., I. Bashir, T. J. Hill S. Taherzadeh, J. Defrance, and P. Jean. 2014. Noise reduction using surface roughness. In *Environmental methods for transport noise reduction*, eds. M. E. Nilsson, R. Klæboe, and J. Bengtsson (pp. 19–47). Oxford, U.K.: Spon Press.
4. Van Renterghem, T., K. Attenborough, M. Maennel, J. Defrance, K. Horoshenkov, J. Kang, I. Bashir, S. Taherzadeh, B. Altreuther, and A. Khan. 2014. Measured light vehicle noise reduction by hedges. *Applied Acoustics* 78: 19–27.
5. Reethof, G., L. Frank, and O. McDaniel. 1977. Sound absorption characteristics of tree bark and forest floor. Paper presented at the *Proceedings of the Conference on Metropolitan Physical Environment*, General Technical Report NE 22: 39–48.
6. Martinez-Sala, R., C. Rubio, L. M. Garcia-Raffi, J. V. Sanchez-Perez, E. A. Sanchez-Perez, and J. Llinares. 2006. Control of noise by trees arranged like sonic crystals. *Journal of Sound and Vibration* 291(1-2): 100–106.
7. Van Renterghem, T., and D. Botteldooren. 2002. Effect of a row of trees behind noise barriers in wind. *Acta Acustica United with Acustica* 88(6): 869–878.
8. Reineke, L. H. 1933. Perfecting stand density index for even-aged stands. *Journal of Agricultural Research* 35: 843–856.
9. Zouboff, V., Y. Brunet, M. Berengier, and E. Sechet. 1994. Paper presented at the *Proceedings of the 6th International Symposium on Long Range Sound Propagation*, June 12–14, Ottawa, Canada (unpublished).
10. DeJong, R., and E. Stusnick. 1976. Scale model studies of the effect of wind on acoustic barrier performance. *Noise Control Engineering* 6: 101–109.
11. Rasmussen, K. B., and M. G. Arranz. 1998. The insertion loss of screens under the influence of wind. *Journal of the Acoustical Society of America* 104(5): 2692–2698.
12. Salomons, E. M. 1999. Reduction of the performance of a noise screen due to screen-induced wind-speed gradients. Numerical computations and wind-tunnel experiments. *Journal of the Acoustical Society of America* 105(4): 2287–2293.
13. Van Renterghem, T., D. Botteldooren, W. Cornelis, and D. Gabriels. 2002. Reducing screen-induced refraction of noise barriers in wind by vegetative screens. *Acta Acustica United with Acustica* 88(2): 231–238.
14. Van Renterghem, T., and D. Botteldooren. 2003. Numerical simulation of the effect of trees on downwind noise barrier performance. *Acta Acustica United with Acustica* 89(5): 764–778.
15. Wilson, J. D. 1985. Numerical-studies of flow through a windbreak. *Journal of Wind Engineering and Industrial Aerodynamics* 21(2): 119–154.

16. Hagen, L. J., and E. L. Skidmore. 1971. Windbreak drag as influenced by porosity. *Transactions of the American Society of Agricultural Engineers* 14(3): 464–465.

17. Van Renterghem, T., and D. Botteldooren. 2013. Designing canopies to improve downwind shielding at various barrier configurations at short and long distance. *Proceedings of the 21th international congress on acoustics* (ICA 2013), Montreal, Canada.

18. Van Renterghem, T., and D. Botteldooren. 2012. On the choice between walls and berms for road traffic noise shielding including wind effects. *Landscape Urban Plan* 105(3): 199–210.

Chapter 6

Noise reduction using surface roughness

Keith Attenborough, Imran Bashir, Toby J. Hill, Shahram Taherzadeh, Jérôme Defrance, and Philippe Jean

Abstract. Sound reflected via the ground interferes with and thereby cancels or reinforces sound travelling in a direct line between source and receiver. The type of interference depends on frequency, the source and receiver locations, and the acoustical properties of the ground surface. For low source heights, the cancellation type of interferences near a hard, smooth surface occurs at such high frequencies that they are ineffective in reducing road and rail noise that affect pedestrians or first- and second-floor residents between 10 and 50 m from the noise source. However, by generating roughness or introducing roughness elements up to 30 cm in height, a more useful range of frequencies is affected, leading to reduced noise. Random spacing of roughness elements is more effective than regular spacing for the broadband noise from road traffic. Use of roughness elements of varying height and clustering can generate larger noise reductions than uniform layouts. Lattice configurations are more effective than the same width of parallel walls because they maintain their effect also when the angle between moving source and receiver deviates from 90°. Roughness treatments using recessed elements are 3 dB less effective than when using protruding elements of the same height. Surface roughness also improves the acoustical performance of a smooth and hard berm. The acoustical performance of surface roughness treatments is improved if the roughness elements are slightly absorbing. Under acoustically neutral meteorological circumstances, the roughness-induced attenuation persists as the receiver moves farther away and if a path is made through the treatment. Conventional fence-type noise barriers lack these desirable features. As is the case with a conventional barrier, the attenuation due to roughness is reduced under downwind and temperature inversion conditions.

6.1 GROUND EFFECT AND ITS MODIFICATION BY ROUGHNESS

6.1.1 Some results of outdoor experiments

Figure 6.1(a) shows nine 7-m-long, 0.2-m-high parallel walls constructed from 594 standard house bricks on an asphalt-covered car park area at

Figure 6.1 (a) Nine low brick walls formed by stacking 594 standard U.K. house bricks on a car park at the Open University, United Kingdom. (b) 3D chequerboard arrangement obtained by displacing pairs of bricks from the parallel walls shown in (a) to halfway between the original walls. (c) Aerial picture of the location of wall array at a distance of between 130 and 140 m from a busy road (Groveway), which is between 5 and 7 m above the car park.

the Open University, Milton Keynes, United Kingdom. Each wall in this arrangement contains 2 rows of 33 bricks, 0.064 m wide. The edge-to-edge spacing between the walls is 0.25 m giving a total array width of 2.57 m. The bricks have been rearranged also to form a "staggered grid" pattern (Figure 6.1(b)). Both the parallel wall and staggered grid arrangements occupy a total area of about 18 m². The location of the car park, about 140 m from a busy main road (Figure 6.1(c)), made it possible to use two microphones to measure the traffic noise levels before and after erecting the walls. The microphone most distant from the road was located either 0.5 m or 1 m from the nearest wall and at heights of either 0.1 m or 0.3 m. The microphone on the road side was 0.5 m from the nearest wall and 0.1 m above the ground.

Before the walls were erected, the differences in traffic noise levels between two microphone locations (about 5 m apart) were negligible because both were about 140 m from the road. Results of measurements with the walls in place indicate that immediately behind the small array of walls the overall A-weighted traffic noise levels were reduced by between 2.5 and 4.4 dB (Figure 6.2(a)(b) and Table 6.1). According to the *principle of reciprocity,** these noise level reductions also should be obtained if the road traffic source were to be in the receiving microphone position (0.1 m above the ground and 0.5 m or 1 m from the nearest wall on the farther side of the wall array) and the receiver were to be located between 130 and 140 m from the road at a height of between 5 and 7 m.

Before bricks are introduced, the surface of the car park is *acoustically hard*. It reflects incident sound like a mirror. Surfaces act as sound mirrors if they are smooth and nonporous like sealed concrete, hot-rolled asphalt, or the surface of unperturbed water. In the reported experiments, the brick devices/constructions modify the reflection of sound and, hence, alter the

---

* Sound preparation is invariant to exchange of source and receives positions.

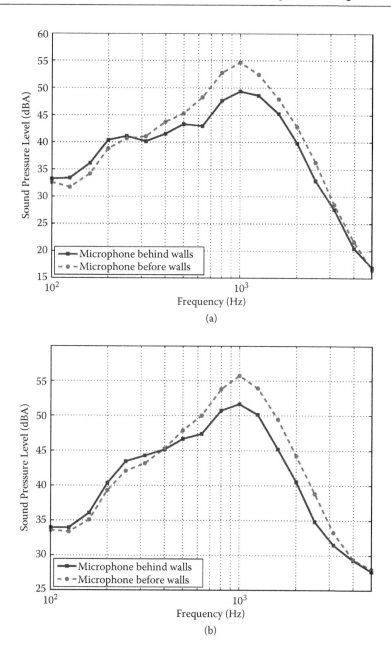

Figure 6.2 Average measured A-weighted sound pressure level spectra before and after the introduction of (a) nine parallel low brick walls (see Figure 6.16(a)) and (b) a "chequerboard" reconfiguration of the bricks (see Figure 6.17(a)–(c)).

Table 6.1 Measured insertion losses (IL) due to the parallel and chequerboard brick walls shown in Figure 6.1(a)(b), respectively, for road traffic noise from a road that is between 130 and 140 m from the wall systems and is elevated by between 5 and 7 m

Distance of source side microphone from nearest wall (m)	Source side microphone height (m)	Average insertion loss for road traffic noise dB	
		2D	3D
0.5	0.1	4.4	3.5
1.0	0.1	4.1	2.9
1.0	0.3		2.5

interaction between the reflected sound and the sound travelling directly from source to receiver. This interaction is known as *ground effect*.[1]

6.1.2 Ground effect as an interference phenomenon

Ground effect involves both cancellation, known as *destructive interference*, and enhancement, known as *constructive interference*. Over smooth, acoustically hard ground, the frequencies at which cancellations and reinforcements occur depend only on the difference between ground-reflected and direct path lengths. The size of this difference is derived simply from source and receiver heights and the distance separating source and receiver positions. The fractional number of wavelengths of the incident sound that fits this path length difference determines the phase angle difference between the direct and ground-reflected sound waves. Cancellation will occur whenever the direct and ground-reflected waves are exactly "out of phase," which is the case when the phase angle difference is an odd integer multiple of 180°. This requires the path length difference to be an odd number of half wavelengths. The lower the source and receiver heights for a given source–receiver separation, the smaller will be the path length difference and the higher will be the frequencies at which cancellation will occur. The direct and ground-reflected sounds reinforce each other when the path length difference is less than half a wavelength or an integer number of wavelengths. This can result in doubling of the incident pressure. In practice, cancellations are never complete because of wavefront spreading, air absorption, and atmospheric turbulence.[*]

If the intervening ground is porous or rough, the interaction between direct and reflected sound depends not only on the difference in the lengths of the direct and ground-reflected paths, but also on the acoustical properties of the ground. Most natural ground surfaces, such as grassland, soil, snow, or gravel, are porous. Sound from a source above the ground is able to penetrate into the pores where there will be friction and thermal exchanges between

---

[*] However, reinforcement may be more or less complete at low frequencies.

the moving air particles and the more or less stationary solid particles. These interactions cause the penetrating sound to change in phase and the resulting phase change adds to that as a result of path length difference. For typical surface transport noise sources and nearby listeners over typical grassland, the first cancellation or destructive interference lies between 300 and 1,000 Hz and may cause a measurable reduction in outdoor noise levels, which is taken into account in noise prediction schemes.[1] The exploitation of ground effect over porous surfaces is discussed in detail in Chapter 7.[2*]

The ground effect for a receiver point is represented by the *excess attenuation* spectrum. The excess attenuation (EA) is the total sound level when there is a ground effect relative to the free field (no ground) level in dB. It represents the "extra" attenuation caused by the ground over and above that following simply from the separation distance and air absorption. At the range of distances and the sound frequencies of interest in this chapter, the small absorption of sound energy by the propagation medium is not important. If source and receiver are placed not too far above an acoustically hard ground surface, the sound pressure level at the receiver is doubled over a wide range of frequencies. In this situation, there is a 6-dB enhancement rather than attenuation of the sound energy. For typical source and receiver heights of interest in surface transport noise problems, the first destructive interference occurs at fairly high frequencies. Consider, for example, point sources at heights of 0.01 m and 0.3 m (these are the heights of the "tyre/road" and "engine" sources on a car according to HARMONOISE.[3] Consider also a listener at 1.5 m or 4 m height separated from the sources by a 20-m acoustically hard surface. For the higher source and lower receiver heights (0.3 m and 1.5 m, respectively), the path length difference is 0.045 m. The first destructive interference will occur when the sound wavelength is twice this path length difference, i.e., 0.09 m. With a sound speed of 340 m/s this affects the frequency given by 340/0.09 = 3,789 Hz. For a receiver at 4 m height and "engine" source, the path length difference is 0.118 m and the first destructive interference occurs at a frequency of 1,445 Hz. The corresponding first cancellation frequencies for the "tyre/road" source are just over 11 kHz and 4 kHz for the 1.5-m receiver and 4-m receiver, respectively.

Figure 6.3 shows the predicted excess attenuation spectra over a smooth, hard surface for "engine" source height (0.3 m) and either 1.5-m- or 4-m-high receivers at a horizontal distance of 20 m from the source. In accordance with the usual convention for such graphs, the excess attenuation (resultant sound level regarding free field) is shown as negative. Negative values signify extra attenuation, whereas positive values signify levels higher than if the ground were absent. The normalised combined source spectrum for a car travelling at 70 km/h[3] also is shown in Figure 6.3.

---

[*] Most of the A-weighted noise made by a car travelling at 70 km/h is in the frequency range between 630 Hz and 4 kHz.[2]

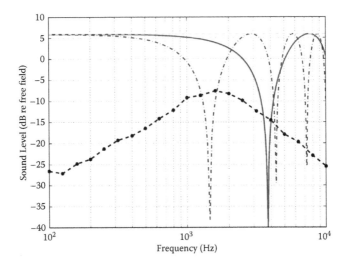

Figure 6.3 Spectrum of sound level regarding free field (excess attenuation) predicted at receivers 1.5 m (solid line) and 4 m (dot–dash line) above an acoustically hard surface and at a horizontal distance of 20 m from a point source at 0.3 m height above the hard surface. A constant +6 DB corresponding to pressure doubling in represented by the horizontal dashed line. Also shown is the third-octave band spectrum (circle-dashed line) for the combined sources on a HARMONOISE[3] car normalised to give an overall sound level of 0 dB.

For the lower receiver and the tyre/road source, the EA spectrum is predicted to be +6 dB up to 10 kHz, which means that the sound pressure is doubled over much of a typical car noise spectrum. So, at a horizontal distance of 20 m from a car travelling at 70 km/h, the sound level due to tyre/road noise would be 6 dB higher in the presence of an acoustically hard ground than it would be above perfectly sound-absorbing ground for both the lower and upper receiver positions. For the engine source and lower receiver, the first destructive interference is predicted to occur at 3.8 kHz and, thereby, influences the sound level from a 70-km/h car between 2 and 6 kHz. More significantly, for the 0.3-m-high source and 4-m-high receiver, the first and second destructive interferences at 1,445 Hz and 4.3 kHz (see Figure 6.1) will have an influence between 800 Hz and 6 kHz on the sound level due to the engine of a car moving at 70 km/h. As a result, for the upper receiver, acoustically hard ground is predicted to increase levels by less than 6 dB. The increase is predicted to be 3.5 dB for the lower receiver and 2 dB for the higher receiver, compared with the level that would be obtained above perfectly absorbing ground. The turbulence that occurs outdoors also tends to decrease the extent of constructive and destructive interference. In an empirical outdoor sound prediction scheme,[4] hard ground is predicted to increase levels by 3 dB.

If a surface is artificially or naturally rough, incident sound is not reflected perfectly, but is scattered by the roughness. The distribution of the scattered sound depends on the roughness topology, the ratio of the roughness dimensions to the incident wavelength, and the relative locations of source and receiver.[1] As long as a sufficient fraction of the incident sound is reflected so that it retains a phase relationship with the incident sound (*coherent reflection*), there can be a significant change in ground effect. When the reflected sound arrives, having opposite phase to that of the direct sound, the ground effect results in a significant extra attenuation for surface transport noise.

In this chapter, a distinction is drawn between topographical features above 0.3 m and those not exceeding 0.3 m in height. Surface features not exceeding 0.3 m are regarded as a form of roughness, whereas features with heights greater than 0.3 m are regarded as barriers (discussed in Chapter 2).[5] The present chapter describes how artificial roughness (≤0.3 m height) can be used to alter the ground effect and exploited for surface transport noise control.

When sound encounters a conventional noise barrier, 2 m or more in height, the top of the barrier acts as a new source, thereby increasing the mean height of the sound path to the receiver. This means that ground effect is less than if the barrier is not present. When the ground on either side of the barrier is acoustically soft, the difference between the sound levels without and with the barrier (known as barrier *insertion loss*) is less than the reduction due to barrier diffraction.

Barriers are most often opaque, shutting out the visual impression of the surroundings. Ground roughness offers an alternative way of controlling noise levels from surface transport without intercepting the line of sight between source and receiver. This will be particularly useful if the land-take and construction costs of the noise-reducing surface are comparable to or lower than those of a conventional barrier, and if pedestrian access across the noise-reducing measure is useful.

Design proposals for reducing noise by rough surface configurations are based on studies of how different layouts affect the EA. Using both measurements and simulations, important aspects are revealed about the way sound travels over rough surfaces and the influence of the various roughness characteristics.

6.2 LABORATORY DATA

6.2.1 Laboratory measurements

The influence of surface roughness on ground effect has been investigated in a systematic manner through measurements made in an anechoic chamber (a laboratory in which the walls, ceiling, and floor are covered with absorbing foam wedges so that very little sound is reflected from them).

Figure 6.4(a) shows a typical measurement configuration. A point source of sound, which is the end of a pipe attached to a Tannoy® loudspeaker, is placed so that the centre of the pipe is at a known height on one side of an array of 19 1-m-long parallel varnished rectangular strips of medium density fibreboard (MDF) that have been spaced regularly across a glass sheet. On the other side of the array is a Bruel and Kjaer half-inch (13-mm) microphone positioned at the same height as the centre of the source pipe.

Figure 6.4(b) shows the results of a series of excess attenuation measurements with 1-m-long varnished wooden, semicylinders of (approximately) 1 cm radius. The source is located 7 cm above a glass plate surface (i.e., a smooth, acoustically hard surface) and the receiver (microphone) is at a horizontal distance of 70 cm from the source and also at a height of 7 cm above the glass plate. For this geometry, the path length difference is 0.014 m and the expected frequency of the first destructive interference over an acoustically hard, smooth surface is 340 ms^{-1}/0.028 m = 12.3 kHz. The solid black line in Figure 6.4(b) shows the excess attenuation spectrum measured without any strip treatment. The EA values with the glass plate alone are near +6 dB up to about 3 kHz, which indicates that, in this frequency range, the sound pressure at the receiver is double that travelling directly from the source. In this situation, the minimum EA is near 12 kHz, which corresponds to the frequency of the first destructive interference between the sound travelling directly to the receiver and that reflected from an acoustically hard surface. The other curves in Figure 6.4(b) show results of increasing the number of regularly spaced semicylinders inserted in parallel on both sides of the dividing line between source and receiver.

The frequency of the destructive interference is reduced from 12.3 kHz to about 11 kHz by placing a single semicylinder on the ground (see the blue dashed line in Fig. 6.4(b)) at the *specular reflection* point that is the point where the hypothetical line from a "mirror image" source to the receiver intersects the ground (Figure 6.5) and is the location at which a sound "ray" passing from source to receiver via the ground is reflected. Since the heights of source and receiver were equal in this experiment, the specular reflection point was halfway between them.

As the number of semicylinders is increased, the frequency of the main destructive interference is progressively lowered and its depth is increased. With 15 regularly spaced semicylinders, the main destructive interference is near 6,500 Hz, i.e., at less than half the frequency of destructive interference for the same source–receiver geometry over the smooth glass plate without any roughness elements present. With nine or more regularly spaced roughness elements, an additional destructive interference near 3,500 Hz becomes noticeable. This is a consequence of the regular spacing of the roughness elements.

Laboratory experiments have been conducted as well with identical parallel strips of various shapes with "random" spacing. A random number generator was used to produce the edge-to-edge separation distances

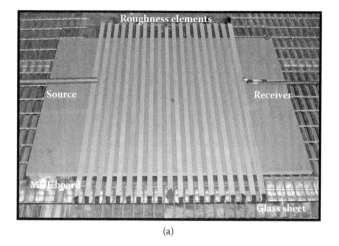

(a)

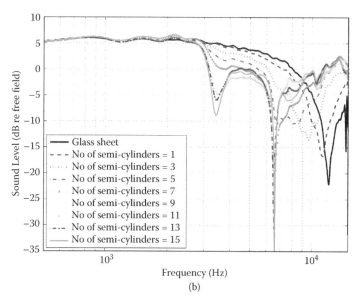

(b)

Figure 6.4 (a) An array of 15 1-cm-radius varnished, wooden semicylinders regularly spaced on an acoustically hard surface (glass plate) in an anechoic chamber. A point source of sound, i.e., the end of a cylindrical tube attached to a Tannoy loudspeaker (not visible), is located at a known height on one side of the array. On the other side of the array is a receiving microphone at the same height as the source. (b) Excess attenuation spectra measured with (point) source and receiver at 0.07 m height and separated by 0.7 m over a glass plate on which are placed between 1 and 15 1-cm-radius varnished, wooden semicylinders with a regular centre-to-centre spacing of 4 cm starting halfway between source and receiver.

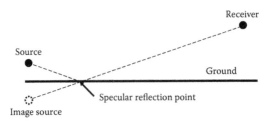

Figure 6.5 Source, image source, receiver, ground, and specular reflection point.

(other than zero). Figure 6.6(a)(b) gives an example of the difference in EA spectra resulting from "random" and regular arrays of the same number of semicylinders for the same source–receiver geometry as for Figure 6.4(a)(b). Each of the random distributions results in a different excess attenuation spectrum. What is shown in Figure 6.6(a)(b) represents the average over five different random distributions. Random roughness distributions result in a single, broad excess attenuation maximum rather than the two or three narrow EA maxima measured when the spacing is regular.

6.2.2 Diffraction-assisted rough ground effect

The extra dips in the EA spectrum associated with regular spacing are the result of *diffraction grating* effects. Diffraction gratings are used in optics. During reflection from a diffraction grating, each of the periodically spaced elements acts as a secondary source of waves that interfere constructively or destructively depending on the angle. At an angle corresponding to constructive interference, the diffracted rays from the adjacent elements will be in phase and the path length difference will be a wavelength or an integer multiple of a wavelength. At all other angles, there will be some destructive interference (cancellation) between the waves. Destructive interference will be at its maximum when the waves are exactly out of phase. For plane wave incident near a grazing incidence, i.e., at an angle of incidence near 90° measured from the normal to the surface, the lowest frequency of maximum destructive interference is given by the speed of sound divided by twice the spacing. For a centre-to-centre roughness element spacing of 5 cm, the lowest diffraction grating interference is at $340/(2 \times 0.05) = 3{,}400$ Hz. The next diffraction grating destructive interference frequency will be at twice this frequency, i.e., 6,800 Hz. These frequencies are close to those of the two dips in the EA spectrum obtained over semicylinders with regular centre-to-centre spacing of 5 cm (see Figure 6.6(a)). However, the lower frequency dip near 3,400 Hz is close also to the dip observed with random spacing and, therefore, may be considered to be a combination of diffraction grating and *roughness-induced* effects. The total excess attenuation spectrum over a periodically rough surface resulting from regular spacing, which includes additional diffraction

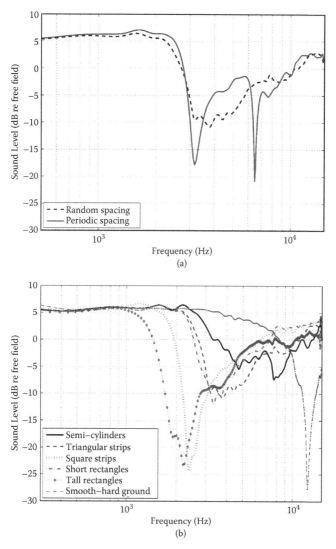

Figure 6.6 Excess attenuation spectra measured with (point) source and receiver at 0.07 m
height and separated by 0.7 m over a glass plate on which are placed (a) 15 1-cm-
radius varnished, wooden semicylinders either randomly with mean centre-to-
centre spacing of 5 cm or regularly distributed at 5 cm spacing about the specular
reflection point (b) randomly distributed semicylinders (black solid line), triangular strips (blue dashed line), square strips (red dotted line), short rectangular
strips (magenta dash-dotted line), or tall rectangular strips (brown dash-diamond
line) with average centre-to-centre spacing of 5 cm. The measured EA spectrum
for the smooth, hard glass base (no roughness) is shown by the brown dotted
(asterisk) curve. The "random" distribution spectra represent the average of
measurements over five different random distributions for each shape.

Table 6.2 Shapes and dimensions of varnished wooden strips used in laboratory experiments

Shape	Height (m)	Width (m)	Cross-sectional area (mm²)
Semicylindrical	0.008	0.02	117
Triangular	0.015	0.029	218
Short rectangular	0.012	0.0285	342
Tall rectangular	0.0285	0.012	342
"Square"	0.018	0.02	360

grating-related dips, can be called *diffraction-assisted ground effect*. Despite the extra peaks introduced by regular spacing, for a given geometry and mean roughness element spacing, the overall broadband insertion losses for random and periodic spacing are similar. On the other hand, regularly spaced roughness might be more practically achievable and aesthetically interesting.

As well as their height and spacing, the shape of the roughness elements is important. Figure 6.6(b) shows example measurements of EA spectra over randomly distributed roughness elements with the shapes listed in Table 6.2. As can be seen from Table 6.2, there were differences in the cross-sectional area of the roughness elements as well as their shape, so the differences in the spectra are not only due to the differences in shape. For a given shape, mean spacing, source location, and receiver location, the taller the roughness elements, the lower the frequency will be of the first destructive interference.

6.2.3 Surface waves: generation and absorption

The excess attenuation spectra measured and predicted over rough surfaces can include values above +6 dB if a *surface wave* is generated. Surface waves are created over surfaces similar to that of a thin porous layer or above a rough hard surface where the roughness features are small compared to the incident wavelengths.[2] A surface wave is a wave that "hugs" the surface with an amplitude that reduces exponentially away from the surface, but at less than the "normal" rate in the horizontal direction. The normal rate of reduction for a point (omnidirectional) source of sound in the absence of any reflecting surfaces is 6 dB per doubling of distance. A surface wave generated by a point source reduces at a rate of only 3 dB per doubling of distance and this poses a problem for roughness-based noise reduction because its decay with distance is not as fast as the decay of other components of the sound field. On the other hand, a substantial proportion of the surface wave energy can be converted into heat, by absorption, thus, the presence of absorption makes it still possible to achieve useful noise reductions with roughness.

Figure 6.7(a)–(d) shows results of measurements in an anechoic chamber during which sound impulses were propagated from a 2-cm-high source over

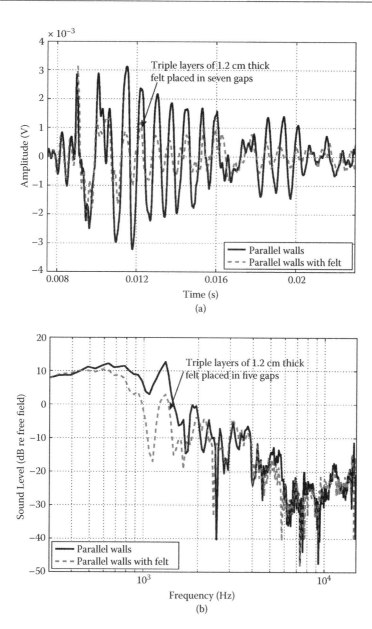

Figure 6.7 (a) Received signals as a function of time and (b) show corresponding excess attenuation spectra above an acoustically hard base supporting 16 4.4-cm-high wooden strips without (solid lines) and with double felt layers (each 1.2 cm thick) placed in the gaps between the strips. The source is 2 cm above the base and 8 cm from the first wooden strip. The receiver is a horizontal distance of 2 m from the source and 10 cm above the base.

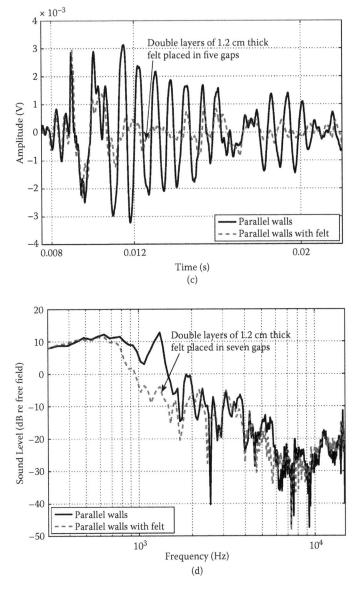

Figure 6.7 (Continued) (c) Received signals as a function of time and (d) corresponding excess attenuation spectra above an acoustically hard base supporting 16 4.4-cm-high wooden strips without (solid lines) and with (dashed lines) double felt layers (each 1.2 cm thick) placed in the gaps between the strips. The source is 2 cm above the base and 8 cm from the first wooden strip. The receiver is a horizontal distance of 2 m from the source and 10 cm above the base.

an acoustically hard wooden base on which were placed 16 4.4-cm-high and 1.5-cm-thick wooden strips with centre-to-centre spacing of 6 cm, starting 8 cm from the source. The receiver was located at a horizontal distance of 2 m from the source and at a height of 10 cm. Figure 6.7(a)–(d) compares the received signals as a function of time obtained without and with any layers of felt placed between some of the wooden strips. In the absence of any felt layers, the first arrival at the receiver has a clear pulse-like waveform. It is followed by a 12-ms-long wave train caused by a surface wave. The period of the surface wave, i.e., the length of time taken for the passage of each complete wave, is 0.7533 ms, which implies that the surface wave frequency (the inverse of the period) is 1,327 Hz.

In the laboratory configurations involving wooden strips, the introduction of materials with good sound absorption at 1,300 Hz, such as layers of felt, reduces the surface wave. This is shown by the data plots in Figure 6.7(a)–(d).

To absorb surface waves outdoors, more practical and robust materials (e.g., soil, gravel, or porous concrete) can be introduced. Typically, the surfaces of outdoor construction materials, such as brick, have a small inherent absorption.

Other ways of reducing surface wave contributions through clusters and height profiles are explored in Section 6.4.

6.3 FIELD DATA FROM BRICK CONFIGURATIONS

6.3.1 Measurements with a loudspeaker

As well as measurements of the influence of the brick wall arrays on road traffic noise (see Section 6.1 and Figure 6.1), measurements have been made also using a Mordaunt loudspeaker placed on the ground at 2 m from the nearest wall of the nine-wall array (see Figure 6.1(a)). The effective source height was 0.1 m. A reference microphone was placed 1 m from the source at a height of 0.1 m. A second microphone was placed successively at 5 m and 10 m from the source at heights 0.21 m, 0.36 m, and 0.85 m. The results of (repeated) measurements using broadband noise from the loudspeaker source are listed in Table 6.3 and show overall insertion losses of between 7.6 and 11.5 dB for parallel wall and lattice arrangements and between 7.2 and 8.3 dB for the 3D staggered grid arrangement.

In the measurements of road traffic noise, the sound spectra at the microphone on the road side of the wall systems were compared with simultaneous spectra obtained at a microphone on the other side of the wall systems. Without the wall systems, the spectra at the two microphones were practically identical. The overall insertion losses following the introduction of the brick arrangements are shown in Table 6.4.

Overall, the measured performance of the 3D staggered grid arrangement is slightly worse than the parallel wall arrangement with the same

Table 6.3 Measured insertion losses using *a loudspeaker* source (effective point source height of 0.1 m) and microphone one at 0.1 m height and 1 m from the nearest point on three-wall configurations at five microphone in two locations

Brick array configurations	Separation (m)	Microphone two height (m)	Average insertion loss (dB)
2.57-m-wide and 0.2-m-high parallel wall array; 9 × 0.064-m-thick walls, 0.31 m centre-to-centre spacing	5.0	0.21	11.0
	5.0	0.25	10.8
	10.0	0.36	9.6
	10.0	0.85	8.5
2.57-m-wide and 0.2-m-high "staggered 3D" arrangement; 9 × 0.064-m-thick rows, 0.31 m centre-to-centre spacing	5.0	0.21	8.3
	5.0	0.25	8.5
	10.0	0.36	7.9
	10.0	0.85	7.2
2.3-m-wide and 0.2-m-high lattice arrangement; 0.215 m × 0.151 m cells, 0.064-m-thick walls, 0.28 m and 0.215 m centre-to-centre spacings	5.0	0.21	11.5
	5.0	0.25	11.0
	10.0	0.36	9.5
	10.0	0.85	7.6

Table 6.4 Measured insertion losses due to the parallel and chequerboard brick walls shown in Figure 6.1(a)(b), respectively, for *traffic* noise from a road that is between 130 and 140 m from the wall systems and is elevated by between 5 and 7 m

Distance of source side microphone from nearest wall (m)	Source side microphone height (m)	Average insertion loss for road traffic noise (dB)	
		2D	3D
0.5	0.1	4.4	3.5
1.0	0.1	4.1	2.9
1.0	0.3		2.5

number of bricks per unit area; the insertion loss is about 1 dB less for the road traffic source and between 2 and 3 dB less for the loudspeaker source.

6.3.2 Drive-by tests

Additional bricks have been used to construct longer arrays on a larger car park at the Open University and, thereby, enabling vehicle pass-by measurements. A total of 1,440 bricks were deployed to create either a 2.14-m-wide, 16-m-long parallel wall array (Figure 6.8(a)) or a 1.1-m-wide, 16-m-long brick lattice (Figure 6.8(b)).

Figure 6.9(a)(b) shows the spectra measured at a 1.5-m-high receiver 10 m from the pass-bys of two different cars (car 1 and car 2) before and after

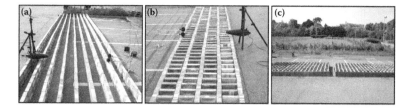

Figure 6.8 Photos of 16-m-long brick configurations used for drive-by measurements, point source loudspeaker, and vertically separated microphones used for effective impedance deduction: (a) 2.14-m-wide array of nine parallel brick walls, (b) a 1.18-m-wide brick lattice, and (c) 2.2-m-wide lattice configuration including a 0.4-m-wide central path.

the insertion of the low wall systems. Each plot represents the average of levels recorded during five pass-bys. The average pass-by speed of car 1 was 36.4 km/h and that of car 2 was 38 km/h. The corresponding broadband insertion losses at a 1.5-m-high receiver 10 m from the nearest wheels were nearly 3 dB. The 1.1-m-wide, square cell lattice wall configuration resulted in more or less the same overall broadband reduction in the noise from car 2 as did the 2.14-m-wide, nine parallel wall configuration for car 1 while occupying significantly less land. The noise reductions measured in the drive-by tests are less than those observed using the distant traffic noise source since, during the drive-by tests, the receiver was 1.5 m high and only 10 m from the source.

Further measurements were made using car 2 to drive past an 8.6-m-long, 1.79-m-wide lattice configuration without and with a central 0.4-m-wide pathway (see Figure 6.8(c)). From the resulting data, averaged insertion losses (three pass-bys at an average speed of 41.0 km/h without a path and four pass-bys at an average speed of 45 km/h with a path) have been calculated. A consequence of the different lattice array dimensions and faster average pass-by speeds is that, without a path, the measured insertion losses (IL) (2.2 dB) are slightly less than that measured (2.6 dB) with the 16-m-long, 1.1-m-wide array. The creation of the path results in a small reduction (about 0.5 dB) in insertion loss.

6.4 PREDICTED EFFECTS OF ROUGHNESS ON ROAD TRAFFIC NOISE

6.4.1 Numerical predictions

The EA spectrum caused by a rough, hard surface can be predicted using numerical methods. Use has been made of a 2D boundary element method (BEM),[6] which involves dividing the surfaces of interest into a number of small hypothetical elements and adding their contributions to the overall sound field resulting from a line source. Although it is a 2D method, it is

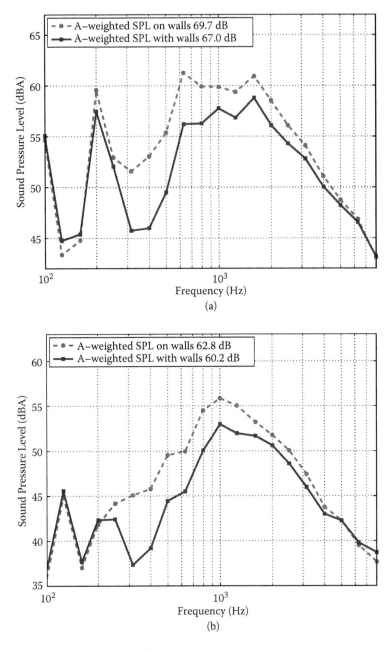

Figure 6.9 Sound pressure level (SPL) spectra measured at the closest point of approach of (a) car 1 and (b) car 2 before and after the insertion of the wall configurations shown in Figure 6.8.

able to predict the sound field from a point source over surfaces containing roughness strips along a line normal to the roughness element axes and has been found to be consistent with laboratory data.[7,8]

The (2D) BEM can be used directly to predict propagation over the parallel wall configurations, but another approach is necessary to enable a 2D BEM to be used to predict propagation over 3D arrangements, such as the lattice. A convenient way of representing the acoustical effects of such surfaces is through their *effective surface impedance*. The surface impedance is the ratio of the pressure at the sound surface to the air particle velocity normal to the surface. The distribution of sound due to a point source over a surface can be predicted from the source and receiver locations and the surface impedance. The impedance of a porous material can be predicted from parameters, such as *flow resistivity* and *porosity*. This is described in more detail in Chapter 7.[2] Conversely, the surface impedance can be deduced from measurements of the received signals due to a point source. Excess attenuation data obtained with point source loudspeaker and vertically separated microphones over a lattice array (see Figure 6.8(b)) have been compared with predictions using a model for the impedance of a slit pore layer with effective flow resistivity = 400 Pa·s·m^{-2}, porosity = 0.55, tortuosity = 1.0, and effective layer depth = 0.25 m. Figure 6.10 compares

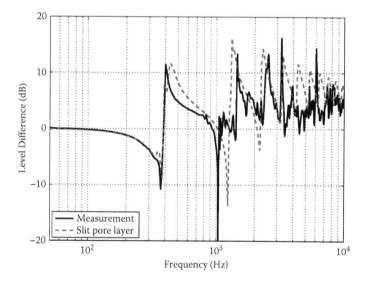

Figure 6.10 Comparison of a measured level difference spectrum (source height = 0.1 m, upper microphone height = 0.15 m, lower microphone height = 0.05 m, horizontal separation = 2 m) above a 0.2-m-high brick lattice with the spectrum predicted by assuming a point source above a slit pore layer impedance with flow resistivity = 400 Pa s m^{-2}, porosity = 0.55, tortuosity = 1.0, and (effective) layer depth = 0.16 m.

measured level difference spectra with predictions made using this effective impedance. The good fit between predictions and data for these geometries means that the slit pore layer impedance can be used with the classical formulae for propagation from a point source over a finite impedance boundary[1] to predict propagation over a (continuous) lattice for other geometries.

The BEM has been used to predict the *insertion loss* caused by roughness configurations parallel to a road, assuming dimensions and geometry intended to be indicative of those that might be used along surface transport corridors. The insertion loss is calculated with respect to a smooth, acoustically hard surface, so it represents the difference in predicted levels over a hard surface before and after the surface is roughened.

6.4.2 Parallel walls versus lattices

A single conventional noise barrier causes sound to *diffract* around its top. The performance of such a barrier depends on the path length difference between the direct path between source and receiver and the path to source and receiver via the top of the barrier. The greater the path length difference, the greater is the attenuation. This means that to be most effective the barrier should be placed as close as possible to source or receiver. If the barrier is positioned close to the source, its effectiveness decreases as the receiver moves farther a way. However, the performance of a parallel low-wall system depends on the interaction of sound with all of the walls. This means that the first wall should not be too close to the source. If it is placed close to the source, diffraction by the first wall dominates the interaction with the wall array. Unlike for a single wall noise barrier, with a source near the ground, the performance of a low parallel wall structure can improve as the receiver moves farther away, but the gain in performance becomes marginal beyond a certain distance. Also, meteorological effects will have increasing importance at longer ranges. Their influence on roughness-based noise barriers will be discussed in a later section.

To make predictions for "standard" two-lane urban roads and four-lane motorways, the European HARMONOISE specifications[3] have been used. Three source heights are specified at 0.01 m, 0.30 m, and 0.75 m. The first height is representative of tyre noise sources and the latter two heights are representative of engine noise sources for cars and other heavier vehicles, respectively. The methodology specifies two standard receiver heights at 1.50 m and 4.00 m intended to be representative of a seated person at ground floor level and at a first floor bedroom window level. The one third octave-band spectrum of a vehicular noise source is specified by the vehicle type (e.g., car, light goods vehicle, or heavy goods vehicle) and vehicle speed. For multiple lanes and a mix of vehicle types, each source location is considered individually and the spectrum weighted according to the traffic flow percentage of each source. Multiple sources are treated as incoherent, i.e., the combination of two identical sources results in a 3-dB increase, as opposed to a 6-dB increase for coherent sources. The predicted EA spectrum and distance attenuation correction is applied to each source and the contribution from each source summed at the receiver location. The insertion loss is calculated with respect to a smooth, acoustically hard surface, i.e., it is the difference in predicted levels before and after the roughness is introduced and takes account, therefore, of excess attenuation due to smooth, hard ground. The wall arrays have been assumed to start 2.5 m from the nearest lane of vehicles. Figure 6.11(a)(b) shows example predictions of sound level spectra.

The predicted insertion losses (IL) of about 6 dB shown in Figure 6.11(a)(b) are reduced if (1) the receiver is closer to the road, (2) the wall height is lower, (3) the array width is smaller, and (4) the receiver is higher. The results listed in Table 6.5 illustrate (1), (2), and (3).

According to Table 6.5, 1-m-wide lattice configurations are predicted to give at least as much insertion loss as parallel wall arrays of twice the width but the same wall spacing perpendicular to the road since their efficacy is less dependent on the azimuthal source–receiver angle than that of the parallel wall array.

The predicted noise reduction due to the proposed deliberate introduction of roughness elements is lower if the proportion of heavy vehicles (which have higher engine noise sources than cars) is greater and if there are traffic lanes farther from the treatment. Nevertheless, at 45 m from the edge of a four-lane motorway carrying 85% cars and 15% lorries at a speed of 70 km/h, a 15-m-wide roughness array containing 26 parallel walls with equilateral triangular cross sections 0.247 m high starting 1 m from the nearside road edge is predicted to give noise reductions of 8.3 dB and 3.2 dB for receivers at heights of 1.5 m and 4 m, respectively. Although there is some effect due to the cross-sectional shape of the roughness elements, the predicted increase in noise reduction that would result from using equilateral triangular wedges rather than 0.3-m-high rectangular wall cross sections (with the same cross-sectional area) for this motorway case is less than 1 dB.

As far as the overall reduction of traffic noise is concerned, there is no clear advantage in the use of periodic rather than random spacing of identical low walls. However, periodic arrangements may be preferred for aesthetic or practical reasons.

Other predictions (not given here) show that a minimum of eight walls is needed to offer a greater insertion loss than a thick wall with the same envelope. Calculations show also that increasing the number of acoustically hard parallel walls from 16 to 30 and, therefore, the width of the array from 3.1 m to 6 m reduces the insertion loss as a result of surface wave creation. On the other hand, the surface wave is decreased by the fact that real brick surfaces are slightly absorbing and can be reduced further by additional absorption, for example, by using a thin layer of gravel between the walls.

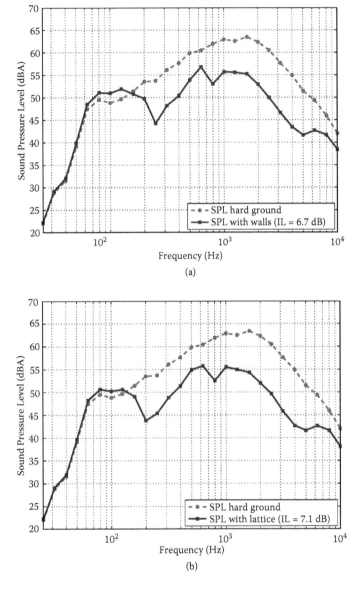

Figure 6.11 Sound level spectra due to a two-lane urban road (95% cars, 5% lorries travelling at 50 km/h) predicted at a 1.5-m-high receiver at a distance of 50 m along the centre line of the array: (a) without and with a 0.3-m-high, low parallel wall array consisting of 16 identical 0.05-m-thick walls with centre-to-centre spacing of 0.2 m (total width 3.05 m); and (b) without and with a 0.3-m-high, 3.05-m-wide lattice (similar to that shown in Figure 1(b)). The arrays are assumed to start 2.5 m from the nearest traffic lane.

Table 6.5 Insertion losses predicted for various roughness configurations and 1.5-m-high receiver locations either 10 m or 50 m from a two-lane urban road

Roughness configurations	Receiver distance (m)	Insertion loss (dB)		
		Lane 1	Lane 2	Combined
3.05-m-wide and 0.3-m-high parallel wall array; 16 × 0.05-m-thick walls, 0.2 m centre-to-centre spacing	50	7.0	6.4	6.7
3.05-m-wide and 0.2-m-high parallel wall array; 16 × 0.05-m-thick walls, 0.2 m centre-to-centre spacing	50	6.6	6.2	6.4
2.14-m-wide and 0.3-m-high parallel wall array; 9 × 0.06-m-thick walls, 0.26 m centre-to-centre spacing	10	3.3	1.4	2.3
2.14-m-wide array and 0.2-m-high parallel wall array; 9 × 0.06-m-thick walls, 0.26 m centre-to-centre spacing	10	2.1	0.4	1.2
3.05-m-wide and 0.3-m-high square lattice	50	6.4	6.0	6.2
3.05-m-wide and 0.2-m-high square lattice	50	6.0	4.8	4.9
2.14-m-wide and 0.3-m-high square lattice	10	3.4	1.4	2.5
2.14-m-wide and 0.2-m-high square lattice	10	2.2	0.7	1.5
1.53-m-wide and 0.3-m-high square lattice	50	6.7	6.2	6.2
1.53-m-wide and 0.2-m-high square lattice	50	4.3	4.0	4.2
1.07-m-wide and 0.3-m-high square lattice	10	3.9	1.9	2.9
1.07-m-wide and 0.2-m-high square lattice	10	2.7	1.0	1.9

With a large number of walls, the insertion loss can be increased by varying the wall height and by using "clusters" of walls. This is discussed in the next section.

6.4.3 Height profiles and clusters

The main predicted influence of a cosine height variation with a period of four walls with centre-to-centre spacing of 0.2 m along a 30-wall array is to introduce an additional EA maximum below 200 Hz. An $N = 17$ quadratic residue diffuser (QRD)[9] profile is predicted also to introduce an additional low frequency excess attenuation maximum, but the main predicted influence of the QRD profile is to smooth the excess attenuation spectrum by reducing both the dips and peaks at higher frequencies. Although there might be some advantage for noise spectra containing lower frequencies, these parallel wall profiles are predicted to offer less than 1 dB increase in the overall insertion loss for car noise compared with 30 identical walls with the same centre-to-centre spacing (0.2 m).

Larger increases in insertion loss for road traffic noise are predicted to result from combinations of fractal profiles and clustering. Figure 6.12(a)–(c)

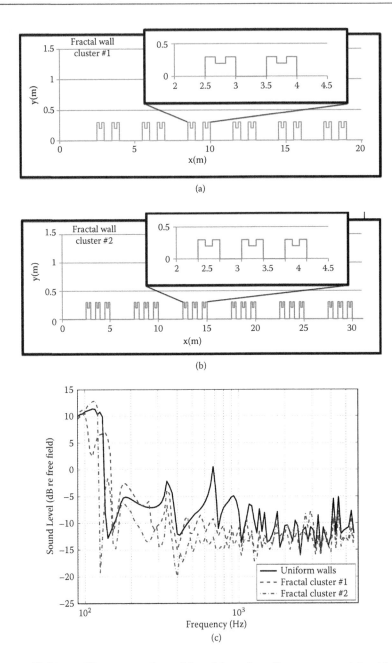

Figure 6.12 "Six wall" portions of two 30-wall fractal profile arrays containing (a) two wall clusters, (b) three wall clusters, and (c) corresponding predictions of excess attenuation spectra compared with that for a 30-m-wide array of identical walls.

shows parts of two example fractal wall clusters and the corresponding excess attenuation predictions. Periodic clustering introduces additional low frequency attenuation, but realisation of these profiles requires larger array widths than considered previously; widths of 16.5 m and 26.5 m rather than 6 m. Fractal arrangement #1 is predicted to yield 2.5 dB higher insertion loss than a uniform wall array with the same width and centre-to-centre spacing (1 m).

6.4.4 Grooves and recessed lattices

Numerical calculations have been made to compare the acoustical performances of raised lattices with equivalent recessed systems having identical "roughness" dimensions (Figure 6.13(a)). Figure 6.13(b) compares the predicted excess attenuation spectrum with 0.1-m-high source and 1.5-m-high receiver separated by 50 m above acoustically hard ground either supporting a 0.3-m-high, 3-m-wide raised lattice or containing a recessed version of the same lattice. The lattices are assumed to begin 2.5 m from the source. The excess attenuation due to the recessed lattice is predicted to be less than that of the corresponding raised case above 450 Hz.

Figure 6.14 and Table 6.6 compare the results of calculations of the insertion losses at a receiver 1.5 m high and 50 m from a two-lane urban road

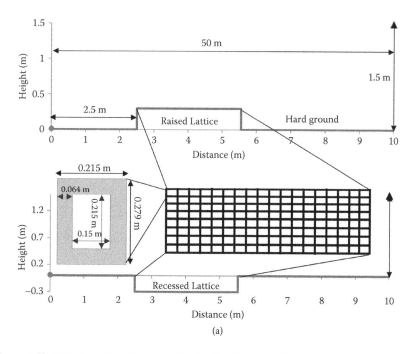

Figure 6.13 (a) Schematic of 0.3-m-high raised and recessed lattice.

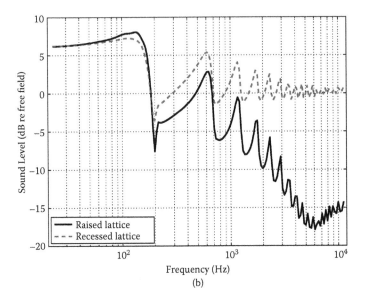

Figure 6.13 (Continued) (b) Comparison between predicted excess attenuation spectra over 3.05-m-wide and 0.3-m-high raised lattice and recessed lattice for a source height of 0.01 m and receiver at a height of 1.5 m and a distance of 50 m from the source. The arrays are assumed to start 2.5 m from the source.

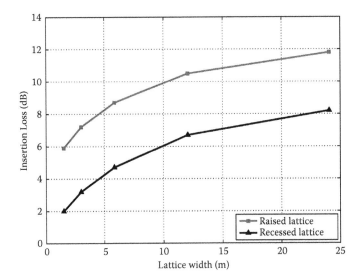

Figure 6.14 Insertion losses as a function of lattice width calculated at a receiver 50 m from a two-lane urban road due to 0.3-m-high (raised) and 0.3-m-deep (recessed) configurations starting 2.5 m from the road.

Table 6.6 Insertion losses (frequency range: 25 — $\frac{1}{n}$ 10 kHz) due to 0.3-m-high lattices (raised or recessed) at a 1.5-m receiver either at approximately 50 m or 100 m from a two-lane urban road

Lattice width	Receiver distance "R" (m)	Receiver height "Hr" (m)	Insertion loss (dB): Raised			Insertion loss (dB): Recessed		
			Lane 1	Lane 2	Combined	Lane 1	Lane 2	Combined
1.53 m	49.5	1.5	6.2	6.5	6.9	2.2	1.9	2.0
		4.0	6.4	4.9	6.6	2.4	1.7	2.0
	102.5	1.5	6.6	6.0	6.3	2.0	1.8	1.9
		4.0	6.3	6.4	6.8	2.2	1.9	2.0
3.05 m	49.5	1.5	7.6	6.7	7.2	3.3	3.0	3.2
		4.0	7.0	6.3	6.1	3.3	2.3	2.8
	102.5	1.5	6.9	6.2	6.5	3.0	2.8	2.9
		4.0	7.5	6.5	7.0	3.3	2.9	3.1
6.85 m	49.5	1.5	9.1	8.2	8.7	4.9	4.4	4.7
		4.0	7.2	6.4	6.3	3.9	2.6	3.2
	102.5	1.5	8.4	7.6	8.0	4.4	4.1	4.2
		4.0	8.8	7.7	8.3	4.8	4.2	4.5
12.05 m	49.5	1.5	10.9	10.2	10.5	7.0	6.4	6.7
		4.0	7.1	6.3	6.1	3.6	2.2	2.9
	102.5	1.5	10.4	9.6	10.0	6.5	6.0	6.3
		4.0	10.2	9.2	9.6	6.5	6.7	6.1
24.05 m	49.5	1.5	12.0	11.6	11.8	8.4	7.9	8.2
		4.0	6.8	6.0	6.9	3.2	1.9	2.5
	102.5	1.5	12.5	11.8	12.1	9.0	8.4	8.7
		4.0	10.9	10.0	10.4	7.4	6.6	7.0

(95% cars, 5% lorries travelling at 50 km/h) due to raised, i.e., 0.3-m-high, and recessed, i.e., 0.3-m-deep, brick lattices starting 2.5 m from the road and with various widths. Typically, the insertion loss is predicted to be between 3 and 4 dB(A) higher for the raised configurations than for the equivalent recessed ones.

While recessed systems are predicted to be acoustically less effective and, potentially, they are more expensive to construct, they might be preferred where there are restrictions on aboveground construction close to roads or where they might be combined usefully with drainage arrangements. It is possible to recover some of the reduced insertion loss by starting them closer to the noise source or by making the recessed configurations deeper than 0.3 m. Being closer to the source also makes roughness-based noise reduction less susceptible to meteorological effects (see Section 6.7).

6.5 PREDICTED EFFECTS OF ROUGHNESS CONFIGURATIONS AROUND RAILWAYS

Figure 6.15 shows potentially noise-reducing arrangements of 0.3-m-high parallel walls (or lattice arrangements) near railway tracks. In the first arrangement, a 3.5-m-wide array of walls is placed on the receiver side starting 1 m from the nearest foot of the ballast profile (greater detail of the assumed track profile and railway noise spectrum are given in Chapter 7).[2] In the second arrangement, 1.5-m-wide arrays of walls are placed between the tracks and on the receiver side. Table 6.7 gives some predicted insertion losses. For example, a 3.05-m-wide configuration of 16 parallel walls starting 1 m from the nearest track is predicted to reduce railway noise

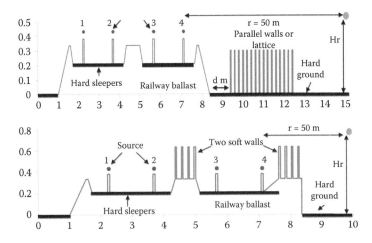

Figure 6.15 Two arrangements of parallel walls or lattices near a two-track railway.

Table 6.7 Insertion loss predictions at a 1.5-m-high receiver due to 0.3-m-high lattices starting 5 m from the nearest railway track

Lattice width	Range "R" (m)	Receiver height "Hr" (m)	Insertion loss (dB): frequency range: 25–10 kHz					
			Source 1	Source 2	Combined 1 & 2	Source 3	Source 4	Combined 3 & 4
1.53 m	49.5	1.5	6.0	4.7	6.5	3.3	3.0	3.2
		4.0	2.1	2.2	2.1	0.5	0.7	0.6
	102.5	1.5	6.5	4.3	6.0	2.9	2.7	2.8
		4.0	6.7	4.4	6.2	2.8	2.6	2.7
3.05 m	49.5	1.5	7.1	6.7	6.6	4.3	4.0	4.2
		4.0	2.1	2.4	2.3	0.6	0.8	0.7
	102.5	1.5	6.5	6.2	6.0	3.8	3.5	3.7
		4.0	6.7	6.3	6.2	3.6	3.4	3.5
6.85 m	49.5	1.5	8.5	7.0	7.9	6.4	6.1	6.3
		4.0	2.1	2.4	2.2	0.5	0.7	0.6
	102.5	1.5	7.9	6.4	7.3	4.9	4.5	4.7
		4.0	7.7	6.4	7.2	4.3	4.2	4.3
12.05 m	49.5	1.5	10.2	8.7	9.6	7.0	6.8	6.9
		4.0	1.8	1.9	1.8	0.2	0.4	0.3
	102.5	1.5	9.7	8.0	9.0	6.7	6.1	6.4
		4.0	8.7	7.6	8.3	6.1	6.2	6.1
24.05 m	49.5	1.5	11.1	10.2	10.7	8.0	8.2	8.1
		4.0	1.6	1.7	1.7	0.0	0.3	0.2
	102.5	1.5	11.5	9.6	10.7	8.6	7.9	8.3
		4.0	8.9	8.3	8.7	6.3	6.7	6.5

by more than 6 dB(A) at a 1.5-m-high receiver 50 m from the edge of the track. A configuration of two 1.5-m-wide, four-wall clusters near the rails is predicted to reduce railway noise by 6.5 dB(A) if the configuration consists of acoustically hard walls, and by 7.7 dB(A) if the configuration consists of slightly absorbing walls. The greater height of railway noise sources compared, for example, with road/tyre contact means that a 0.3-m-high roughness arrangement gives higher insertion losses if placed farther away from the track.

6.6 PREDICTED EFFECTS OF SURFACE ROUGHNESS ON THE ACOUSTICAL PERFORMANCE OF BERMS

Sometimes bunds (embankments) or berms are preferred to fence-type barriers for noise control. If the berms are constructed with compacted soil surfaces, the deliberate introduction of surface roughness in the form of parallel walls or grooves can improve their acoustical performance. Figure 6.16(a) shows an example of a trapezoidal berm next to a four-lane motorway (85% cars and 15% HGV (heavy goods vehicle) mean speed 70 km/h).

Averaged over the receiver zone indicated in Figure 6.16(a), the smooth, acoustically hard berm is predicted to yield 18 dB traffic noise reduction. Nevertheless, even this substantial reduction can be improved by the construction of narrow 0.2-m-high parallel grooves (Figure 6.16(b)) on the top of the berm. This measure is predicted to give a further 7 dB noise reduction.

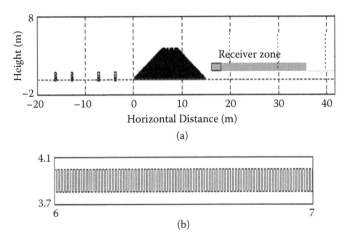

Figure 6.16 (a) Schematic of a berm (15-m-wide base, 3-m-wide top, and 4 m high) next to a four-lane motorway. The receiver zone is between 1 and 2 m above the ground plane and extends for 20 m starting 1 m beyond the receiver-side edge of the berm. (b) Detail of a 1-m-wide section of thin parallel grooves (0.2 m deep, 1.25 cm wide, 2.5 cm centre-to-centre spacing) at the top of the berm shown in (a).

This is equivalent to the noise reduction that would result from a 1-m increase in the height of the smooth berm.

For the berm cross section in Figure 6.16(a), additional noise reduction (up to 3 dB) is predicted if all three sides of the berm are roughened in the same manner or with variable height roughness. Although specific berm-related calculations have not been made for lattice roughness, the evidence from the calculations reported earlier is that a lattice roughness arrangement should be even more effective per unit array width and less azimuthal angle dependent than parallel grooves.

6.7 METEOROLOGICAL EFFECTS ON ROUGHNESS-BASED NOISE REDUCTION

Downward refracting conditions are known to have a deleterious influence on the acoustical performance of fence-type noise barriers. Such conditions may be expected also to have an adverse influence on the noise reduction due to ground treatments. The influence of downwind conditions on the excess attenuation due to (1) a 3-m-wide, 0.26-m-deep recessed lattice treatment starting 0.5 m from a 0.01-m-high source (road/tyre); and (2) a 3-m-wide, 0.26-m-high lattice starting 2.5 m from the source has been studied using a 2D numerical parabolic equation prediction scheme,[10] which assumes a line source. The lattice surface is modelled like that of a slit-pore layer (see Chapter 7)[2] with effective flow resistivity = 400 Pa s m^{-2}, porosity = 0.55, tortuosity = 1.0, and effective layer depth = 0.25 m. The atmosphere is assumed to have a logarithmic sound speed profile given by:

$$c(z) = c_0 + b1n\left(1 + \frac{z}{z_0}\right) \tag{6.1}$$

with $b = 0$ for an acoustically neutral atmosphere and $b = 4$ for downwind. Also allowance has been made for a moderate amount of turbulence (see, for example, Attenborough et al.[1]: outer limit of the length scale = 1.1 m; variance of refractive index = 5×10^{-6}). However, the influence of turbulence is small at the 50-m range.

Figure 6.17(a)(b) shows predictions of excess attenuation spectra for acoustically neutral and downward refracting conditions over a 0.26-m-high, 3-m-wide lattice at a 1.5-m-high receiver, 50 m from the 0.01-m-high source. These predictions indicate that, at a 50-m range, although the excess attenuation in the absence of meteorological effects (the "neutral" curve) is higher overall for the raised lattice, the influence of the downward refraction on the acoustical performance is larger for the raised lattice than for the recessed one particularly near 1 kHz. It should be noted, as shown in Figure 6.17(c), that downward refraction is predicted to have a significantly adverse effect on the excess attenuation spectrum over the recessed lattice at 100 m.

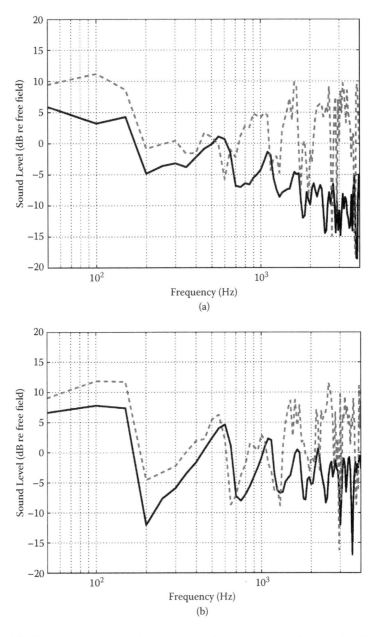

Figure 6.17 EA spectra predicted for acoustically neutral and downward refracting meteorological conditions for a 0.01-m-high (line) source: (a) at a 1.5-m-high receiver 50 m from the source due to a 0.26-m-high, 3-m-wide lattice with nearest edge 2.5 m from the source; (b) at a 1.5-m-high receiver 50 m from the source due to a 0.26-m-deep recessed lattice starting 0.5 m from source.

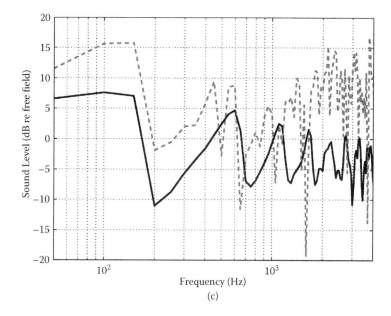

Figure 6.17 (Continued) EA spectra predicted for acoustically neutral and downward refracting meteorological conditions for a 0.01-m-high (line) source: (c) for the recessed lattice described in (b), but at a 1.5-m-high receiver 100 m from the source.

6.8 CONCLUSION

Significant reductions in surface transport noise (up to 10 dB) can be obtained by the deliberate introduction of at least 3-m-wide strips of 0.3-m-high roughness on flat, hard ground. A particularly effective form of roughness has the form of a 0.3-m-high, 0.2-m-side, 3-m-wide square cell lattice that offers greater insertion loss and has less azimuthal angle dependence than a parallel wall array with 0.2 m spacing and 3 m width.

Roughness treatments can be recessed; however, this reduces their insertion loss by about 3 dB. Although ground roughness-based reductions are comparable only with those offered by a relatively low (1-m-high) single barrier and use more land, they might be an attractive alternative to such a barrier where it is desirable to preserve line of sight. Moreover, the performance of a roughness treatment is not reduced significantly if a path is made through it. Also, unlike a conventional noise barrier, the acoustical performance of a rough ground treatment increases as the source–receiver distance is increased. In common with a conventional noise barrier, the acoustical performance of a roughness-based noise barrier is reduced by downward refracting and turbulent meteorological conditions.

REFERENCES

1. Attenborough, K., K. M. Li, and K. Horoshenkov. 2007. *Predicting outdoor sound*. London: Taylor & Francis.
2. Attenborough, K., S. Taherzadeh, I. Bashir, J. Forssén, B. Van der Aa, and M. Männel. 2014. Porous ground, crops and buried resonators. In *Environmental methods for transport noise reduction*, eds. M. E. Nilsson, R. Klæboe, and J. Bengtsson (pp. 155–177). Oxford, U.K.: Spon Press.
3. Nota, R., R. Barelds, and D. Van Maercke. 2005. *Harmonoise WP 3 engineering method for road traffic and railway noise after validation and fine tuning*. Deliverable of WP3 of the HARMONOISE project. Document ID HAR32TR-040922-DGMR20 (2005).
4. ISO. 1996. *Acoustics—Attenuation of sound during propagation outdoors—Part 2: A general method of calculation (ISO 9613-2)*. Geneva: International Standards Organization.
5. Defrance, J., P. Jean, F. Koussa, T. Van Renterghem, J. Kang, and J. Smyrnowa. 2014. Innovative barriers. In *Environmental methods for transport noise reduction*, eds. M. E. Nilsson, R. Klæboe, and J. Bengtsson. Oxford, U.K.: Spon Press.
6. Chandler-Wilde, S., and D. Hothersall. 1985. Sound propagation above an inhomogeneous impedance plane. *Journal of Sound and Vibration* 98(4): 475–491.
7. Boulanger, P., K. Attenborough, S. Taherzadeh, T. Waters-Fuller, and K. M. Li. 1998. Ground effect over hard rough surfaces. *Journal of the Acoustical Society of America* 104(3): 1474–1482.
8. Bashir, I., S. Taherzadeh, and K. Attenborough. 2013. Diffraction assisted rough ground effect: Models and data. *Journal of the Acoustical Society of America* 133 (3): 1281–1292.
9. Cox, T.J., and D'Antonio, P. 2004. *Acoustic Absorbers and Diffusers*. London and New York: Spon Press.
10. Salomons, E. 2001. *Computational atmospheric acoustics*. Dordrecht, The Netherlands: Kluwer Academic Publishers.

Chapter 7

Porous ground, crops, and buried resonators

Keith Attenborough, Shahram Taherzadeh, Imran Bashir, Jens Forssén, Bart Van der Aa, and Manuel Männel

Abstract. Transport noise travelling near to naturally occurring porous ground surfaces, such as grassland, is reduced at frequencies relevant to noise control as the result of interference between direct and ground-reflected sound known as the *ground effect*. Although it is observed frequently that different types of naturally occurring ground surfaces give rise to different ground effects, little thought has been given to deliberately choosing ground to give an improved noise reduction. In this chapter, results of calculations and measurements are provided that can inform this choice. Dense vegetation, in the form of ground cover or crops, contributes to ground effect as a result of the penetration of roots into soil and adds extra attenuation due to scattering by leaves and stems. Calculations are provided that suggest the potential usefulness for noise control of certain combinations of ground and crops. Experiments and calculations are reported showing that ground that is otherwise acoustically hard, such as nonporous asphalt or concrete, can be altered to be more effective for noise reduction by being made porous or by inserting porous strips or patches composed, for example, from gravel. Porous asphalt road surfaces are used to reduce both noise generation and propagation from road–tyre interaction. Their effectiveness for reducing traffic noise can be increased by burying preformed resonating chambers. Resonators also can be buried in hard or porous ground to reduce noise. Laboratory data and predictions are presented that demonstrate the usefulness of this method.

7.1 POROUS GROUND AND CROPS

Attenborough et al.[1] (and in this volume, Chapter 6) have described the phenomenon of ground effect over hard and rough surfaces and its representation by excess attenuation spectra. The excess attenuation at a given receiver may be calculated from knowledge of the surface impedance and the source–receiver geometry. Most naturally occurring outdoor surfaces are porous. As a result of being able to penetrate the porous surface, ground-reflected sound is subject to a change in phase as well as having some of

Table 7.1 Categories of ground included in the HARMONOISE and NORD2000 outdoor sound prediction models

Categories	Examples
A. Very soft	Snow or moss-like
B. Soft forest floor	Short, dense heather-like, or thick moss
C. Uncompacted, loose ground	Turf, grass, loose soil
D. Normal uncompacted ground	Forest floor, pasture field
E. Compacted field and gravel	Compacted lawns, park area
F. Compacted dense ground	Gravel road, parking lot
G. Hard surfaces	Dense asphalt, concrete, water

its energy converted into heat. In earlier outdoor noise prediction schemes, ground surfaces were considered as either "acoustically hard," which means that they are perfectly reflecting, or "acoustically soft," which implies that they are perfectly absorbing. According to ISO 9613-2,[2] any ground surface of low porosity is "acoustically hard" and any grass-, tree-, or potentially vegetation-covered ground is "acoustically soft." Although this might be an adequate representation in some circumstances, it oversimplifies a considerable range of properties and resulting effects. Even the category of ground known as *grassland* involves a wide range of ground effects.[3]

Porosity is not the only factor that influences the acoustical properties of porous ground. They are affected most by the ease with which air can move in and out of the ground surface. This is indicated by the *flow resistivity,* which represents the ratio of the applied pressure gradient to the induced volume flow rate of air through the surface of the ground. The porosity of naturally occurring ground surfaces does not vary as much as their flow resistivity. If the ground surface has a high flow resistivity, it means that it is difficult for air to flow through the surface. This can result from very low surface porosity. Hot-rolled asphalt and nonporous concrete have a very high flow resistivity, whereas many forest floors and freshly fallen snow have a much lower flow resistivity. The HARMONOISE prediction scheme[4] identifies eight categories of ground (Table 7.1) for the purposes of predicting the extra sound attenuation associated with ground effect. According to Table 7.1, various types of grass-covered ground span the HARMONOISE categories C, D, and E. However, as is the case with other current prediction schemes, the HARMONOISE procedure does not consider the potential of deliberately choosing a particular ground surface for improved noise control.

7.1.1 Replacing hard ground with soft ground

Figures 7.1 to 7.3 show nine types of grass-covered ground in Sweden, the acoustical properties of which have been determined by fitting data from short-range measurements.[5]

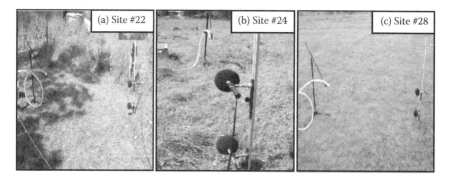

Figure 7.1 Three types of grass-covered ground predicted to give lower traffic noise reductions.

Figure 7.2 Three types of grass-covered ground predicted to give higher traffic noise reductions.

Figure 7.3 Three types of ground with low flow resistivity, but with acoustical properties best described as those of a hard-backed layer.

Best fit values of flow resistivity and porosity are listed in Table 7.2. They were obtained by fitting short-range sound propagation data after assuming various models for the acoustical surface impedance, which represents the ratio of the incident pressure to the induced air particle velocity. The fitted flow resistivity values vary by a factor of 20. On the other hand, with the exception of the sports field, the fitted porosity values are similar.

The fitted parameter values listed in Table 7.2 can be used to predict reductions in noise levels at a distance of 50 m from the closest point of approach on a two-lane urban road (95% cars type 1 and 5% heavy goods vehicles (HGVs)) travelling at a mean speed of 50 km/h (Figure 7.4) that would result from replacing acoustically hard ground surfaces by porous grass-covered surfaces. The discontinuity from hard to soft ground has been taken into account by using a semiempirical method.[6] This method allows for the different acoustical impedance of the hard and soft ground and the source–receiver discontinuity geometry, and has been validated by experiments with a single discontinuity. Three values of the distance x m from the nearest lane to the start of the soft ground have been considered. According to the calculation results listed in Table 7.3, replacing a 45-m-wide strip of acoustically hard ground

Table 7.2 Parameter values for the ground types shown in Figures 7.2 to 7.4 used to calculate the results listed in Tables 7.3, 7.4, and 7.5

Surface description	Porosity	Flow resistivity kPa s m^{-2}	Layer depth m
#22 Pasture	0.5	1344	∞
#24 Arable	0.50	2251	∞
#28 Sports field	0.22	664	∞
#16 Lawn	0.5	176	∞
#18 Arable	0.7	397	∞
#41 Long grass	0.36	104	∞
#7 Arable	0.56	85	0.034
#9 Urban	0.52	59	0.050
#27 Long grass	0.65	51	0.046

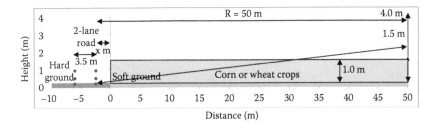

Figure 7.4 Schematic of the configurations used to calculate the noise reductions listed in Tables 7.2, 7.3, 7.4, and 7.5.

Table 7.3 Noise reductions from a two-lane urban road calculated for the configuration shown in Figure 7.4 after replacing hard ground with the three ground types shown in Figure 7.1

| Surface description | x m | Reduction (dB) compared with smooth, hard ground | | | | | |
| | | Hr = 1.5 m | | | Hr = 4 m | | |
		Lane 1	Lane 2	Combined lanes	Lane 1	Lane 2	Combined lanes
#22 Pasture	0	6.4	6.6	6.5	3.7	3.0	3.3
	2.5	6.4	6.3	6.3	3.0	2.2	2.6
	5	6.1	5.9	6.0	2.2	1.5	1.9
#24 Arable	0	5.5	5.7	5.6	5.5	5.7	2.9
	2.5	1.5	5.5	5.5	5.5	5.5	1.7
	5	1.5	5.3	5.3	2.0	1.4	1.3
#28 Sports field	0	6.2	6.2	6.2	3.6	2.8	3.2
	2.5	6.1	5.9	6.0	2.8	2.1	2.4
	5	5.8	5.6	5.7	2.1	1.4	1.7

by any acoustically soft ground *will decrease levels by at least 5 dB at a 1.5-m-high receiver and by between 1 dB and 3.5 dB at a 4-m-high receiver*. Although these calculations have assumed that the receiver is located over soft ground, additional calculations show that a receiver above hard ground and within 5 m of the nearest soft ground would experience similar noise reductions.

The calculation results listed in Table 7.4 suggest that the difference between the types of soft ground illustrated in Figures 7.1 and 7.2 could result in a *nearly 3-dB difference in the sound levels* at a 1.5-m-high receiver 50 m from the road as long as the soft ground extends from 2.5 m from the road edge to the receiver. The difference in predicted noise reductions between ground types is smaller if the distance between the nearest lane and the start of the soft ground is increased to 5 m and/or the receiver height is increased to 4 m.

An important difference between these types of grassland is revealed by the (fitted) flow resistivity values listed in Table 7.2. Higher noise reductions are predicted if the ground has relatively low flow resistivity. Ground surfaces that have been compacted, for example, by frequent mowing, rolling, or heavy wheeling, are likely to have higher flow resistivity.

The predicted reductions in Table 7.5 indicate that, even if the ground has a low flow resistivity, the presence of an acoustically hard layer at a shallow depth, which is implied by the short-range data fitting, will result in a slightly lower overall noise reduction than for other nonlayered grounds with comparably low-flow resistivity. Mature forest floors are likely to

Table 7.4 Noise reductions from a two-lane urban road calculated for the configuration shown in Figure 7.4 after replacing hard ground with the three ground types shown in Figure 7.2

		Reductions (dB) compared with smooth, hard ground					
		Hr = 1.5 m			*Hr = 4 m*		
Surface description	*x m*	*Lane 1*	*Lane 2*	Combined lanes	*Lane 1*	*Lane 2*	Combined lanes
#16 Lawn	0	9.4	9.1	9.3	5.5	3.9	4.6
	2.5	9.0	8.4	8.7	4.1	2.6	3.3
	5	8.3	7.6	7.9	2.8	1.6	2.1
#18 Arable	0	9.0	8.8	8.9	5.2	3.8	4.4
	2.5	8.7	8.2	8.4	4.0	2.7	3.3
	5	8.1	7.5	7.8	2.8	1.7	2.2
#41 Long grass	0	9.5	9.0	9.3	5.6	3.8	4.6
	2.5	8.9	8.3	8.6	4.0	2.5	3.2
	5	8.2	7.4	7.8	2.6	1.5	2.0

Table 7.5 Noise reductions from a two-lane urban road calculated for the configuration shown in Figure 7.4 after replacing hard ground with the three ground types shown in Figure 7.3 using the hard-backed layer parameter values

		Reduction (dB) compared with smooth, hard ground					
		Hr = 1.5 m			*Hr = 4 m*		
Surface description	*x m*	*Lane 1*	*Lane 2*	Combined lanes	*Lane 1*	*Lane 2*	Combined lanes
#7 Arable	0	8.2	8.0	8.1	5.2	3.7	4.4
	2.5	7.9	7.5	7.7	3.9	2.5	3.1
	5	7.4	6.8	7.1	2.6	1.4	2.0
#9 Urban	0	8.6	8.4	8.5	5.4	3.7	4.4
	2.5	8.3	7.7	8.0	3.9	2.4	3.1
	5	7.7	7.0	7.3	2.6	1.3	1.9
#27 Long grass	0	8.6	8.4	8.5	5.2	3.7	4.4
	2.5	8.3	7.7	8.0	3.9	2.4	3.1
	5	7.7	7.0	7.3	2.5	1.3	1.9

have a low-flow resistivity as a result of a layer of partially decomposed vegetation above the soil substrate. Noise reduction by tree belts is considered in detail by Van Renterghem et al.,[7] and in this volume, Chapter 5. However, it should be noted that half of the overall noise reduction that is predicted for a narrow mature tree belt is a consequence of the low-flow resistivity ground effect.

7.1.2 Reduction of tramway noise after replacing asphalt with grass

A measurement campaign in Grenoble, France, has compared tramway noise levels in the situations shown in Figure 7.5.[8] In the left-hand photo there is grass between and alongside the tracks in place of the acoustically hard surfaces shown in these locations in the right-hand photo. Results of noise level measurements in these situations have shown that the introduction of grass between and alongside the tracks reduces tram noise levels at a 1.5-m-high receiver about 4 m from the nearest track by between 1 and 10 dB(A) with an average of about 3 dB(A).

Calculations made using the boundary element method (BEM), assuming vertical arrays of point sources at heights of 0.05 m, 0.3 m, and 0.5 m to represent each wheel as a noise source (Figure 7.6), a standard tram noise spectrum, and grass-type impedance spectra, suggest reductions of between 1 and 6 dBA for trams on the farther track and of between 0.5 and 4.5 dBA for trams on the nearer track (Table 7.6).

Figure 7.5 Contrasting ground surfaces (a) grassland and (b) asphalt around a tramway in grenoble, France.[8] With permission.

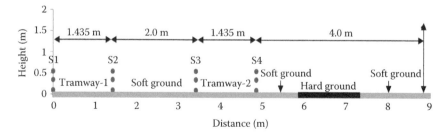

Figure 7.6 Sources and receiver locations assumed for predicting Grenoble tram track noise levels.

Table 7.6 Predicted insertion losses due to the introduction of grass near a tramway at Grenoble, France

Ground parameters		Source height "Hs" (m)	Insertion loss (dB) frequency range: 25–10 kHz					
Flow resistivity (kPa s m⁻²)	Porosity		Source – S1	Source – S2	Combined S1 & S2	Source – S3	Source – S4	Combined S3 & S4
100	0.5	0.05	6.6	5.9	6.3	5.0	4.1	4.5
		0.3	3.0	2.1	2.6	0.8	0.6	0.7
		0.5	0.5	0.8	0.7	1.3	1.1	1.2
200	0.5	0.05	5.1	4.6	4.9	3.9	3.1	3.5
		0.3	3.2	2.3	2.8	1.0	0.4	0.7
		0.5	0.8	0.9	0.9	1.1	0.8	0.9
300	0.5	0.05	4.3	3.9	4.1	3.3	2.5	2.9
		0.3	3.2	2.4	2.8	1.0	0.3	0.6
		0.5	1.0	0.9	0.9	0.9	0.6	0.8
500	0.5	0.05	3.5	3.1	3.3	2.6	2.0	2.3
		0.3	2.9	2.2	2.6	1.0	0.3	0.6
		0.5	1.0	0.9	1.0	0.8	0.5	0.6
1,000	0.5	0.05	2.5	2.2	2.4	1.9	1.4	1.7
		0.3	2.4	1.9	2.2	0.9	0.2	0.5
		0.5	1.0	0.8	0.9	0.6	0.3	0.5
2,000	0.5	0.05	1.8	1.6	1.7	1.4	1.0	1.2
		0.3	1.9	1.5	1.7	0.7	0.1	0.4
		0.5	0.9	0.7	0.8	0.4	0.2	0.3

7.1.3 Replacing a road with hard strips in otherwise soft ground

Replacing a road with 0.3-m-wide, acoustically hard, driving strips in otherwise acoustically soft ground, consisting of low-flow resistivity grass or gravel, is predicted to give insertion losses even greater than those predicted and measured for the introduction of grass between and alongside tramway tracks. Figure 7.7 shows the assumed geometry in which each car on a single lane is idealised as four-point sources (near-side engine and tyre, plus off-side engine and tyre) with a 1.6-mm-wide and 1.2-m-high rectangular, acoustically hard box as the simplified car body. Table 7.7 shows the predicted insertion losses, taking into account the car body, for near-side and off-side pairs of sources and the combined predicted insertion loss when gravel replaces the acoustically hard road surface apart from the 0.3-m-wide driving strips. The predictions listed in Table 7.7 are for infinitely deep gravel, but other calculations show that 0.1 m depth of gravel should be sufficient.

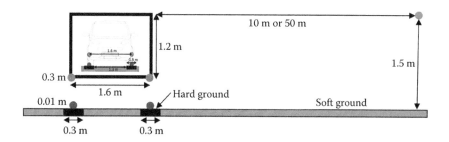

Figure 7.7 Idealised 1.6-m-wide × 1.2-m-high car body "box" and source–receiver locations assumed for BEM calculations.

Table 7.7 Insertion loss (25 Hz to 10 kHz) predicted at the 50-m range from a single lane of cars travelling at 50 km/h on hard 0.3-m-wide driving strips surrounded by at least 0.1 m depth of gravel (flow resistivity 10 kPa^s m^{-2}, porosity 0.5) or at least 0.2 m depth of grass (flow resistivity 104 kPa^s m^{-2}, porosity 0.6)

		Insertion loss dB		
Soft ground type	Receiver height (m)	Near side	Off side	Combined
Gravel	1.5	9.3	18.4	11.6
	4.0	3.8	12.3	6.1
Low flow resistivity grass	1.5	8.4	15.3	10.5
	4.0	4.0	11.3	6.1

7.1.4 Combined effects of crops and ground

Based on an analysis of data for sound transmission loss through various types of vegetation including corn, maize, and reeds,[9,10] it is possible to predict the extra attenuation (i.e., in addition to that due to soft ground effect) due to foliage as a function of frequency and crop characteristics using the empirical formula:

$$\frac{EA(dB)}{\sqrt{FL}} = A\left[1 - \exp(0.3 - 0.5(ka))\right], \qquad ka \geq 0.6, \tag{7.1}$$

where $EA(dB)$ represents the excess attenuation in dB, F/m is the foliage area per unit volume, L m is the length of the propagation path, k is the wavenumber = $2\pi f/c$, c is the adiabatic sound speed in air, and a m is the mean leaf width. A is a constant with a value of 3 for best fit to Aylor's data. The lower limit on ka avoids negative values of EA.

The reductions in noise in dB/m calculated for four combinations of two types of 1-m-high crops with two types of soft ground assuming the configuration depicted in Figure 7.2 with $x = 5$ m are listed in Table 7.8.

Table 7.8 The attenuation (dB) of noise from a two-lane urban road predicted for four combinations of soft ground and 1-m-high crops, assuming the configuration shown in Figure 7.2

		Reduction (dB) compared with smooth, hard ground			
		1.5-m-high receiver		*4-m-high receiver*	
Ground	*Crop*	*Ground alone*	*Ground + crop*	*Ground alone*	*Ground + crop*
#18 Arable	Dense corn	7.8	13.1	2.2	6.7
	Winter wheat		8.6		2.5
#24 Arable	Dense corn	5.3	9.0	1.7	5.4
	Winter wheat		5.6		1.9

The dense corn crop is characterised by a leaf area per unit volume of 6.3/m and a mean leaf size of 0.0784 m. For winter wheat, the corresponding values are 30/m and 0.012 m, i.e., the winter wheat has a higher foliage area per unit volume but smaller leaves. The overall attenuation is calculated as the sum of that due to ground effect and the attenuation along the parts of the direct paths from the vehicle sources to the receivers that pass through the crop (see Figure 7.2). The combination of high-flow resistivity and small leaf crop is predicted to have little acoustical merit. On the other hand, combinations of low-flow resistivity ground and dense large leaf crops are predicted to give a *total attenuation of between 9 and 13 dB* at the 1.5-m-high receiver of which *between 1 and 5 dB is contributed by the crops*. The corresponding predicted total attenuations at the 4-m-high receiver are between 2.5 and 7 dB of which between 0.3 and 4.5 dB is contributed by crops. For the considered configuration, the 1-m-high crop is predicted to offer relatively little additional attenuation at the 4-m-high receiver.

7.1.5 Acoustically soft strips and patches

Laboratory experiments have investigated the effects of acoustically soft strips and patches in an otherwise hard surface. Figure 7.8 shows two examples of experimental configurations in which either felt or sand has been used to provide an acoustically soft material and medium density fibreboard (MDF) has been used to provide an acoustically hard material.

These "strip" and "patch" arrangements give equal areas of "hard" and "soft" ground. As long as the source–receiver axis was perpendicular to, i.e., at right angles to, the parallel strip axes, the resulting data show that both arrangements produce more or less the same ground effect. However, if the source–receiver axis is inclined at greater angles to parallel "strip"

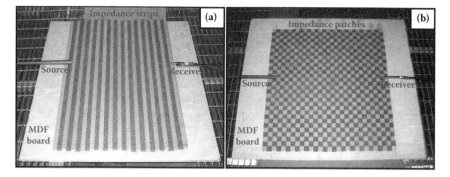

Figure 7.8 Laboratory configurations used to investigate sound propagation over (a) acoustically soft strips and (b) patches of felt. The acoustically hard material is varnished medium density fibreboard (MDF).

axes, the effect changes and the overall A-weighted reduction may be reduced. The change is particularly dramatic for azimuthal angles greater than ±30° from the perpendicular to the strips. On the other hand, the ground effect due to a "patches" arrangement in the laboratory was found to be more or less independent of the azimuthal angle between source and receiver. This suggests that a "patches" arrangement may be preferable for traffic noise control.

Figure 7.9(a)(b) shows two potential configurations next to a two-lane urban road (95% cars, 5% heavy vehicles, 50 km/h) with either a single wide strip or several narrow strips of gravel alternating with equally wide "hard" strips between the road and receivers at 1.5 m and 4 m height and 50 m distance from the edge of the road.

Table 7.9 lists predicted insertion losses for various gravel and grass strip widths and configurations (defined by *a* and *b* in Figure 7.6). The strips are assumed to have acoustical properties that correspond to infinitely thick gravel (with 3 mm mean stone size) or grass (flow resistivity = 150 kPa·s·m^{-2}, porosity = 0.5). Although multiple relatively narrow strips may be preferred for aesthetic or practical reasons, they do not achieve any greater reduction than a single strip of the same total width. Moreover the number of strips within a given area of ground is not predicted to influence the noise reduction much. For the considered geometry and gravel, no increase in insertion loss is predicted if the width of a single strip is increased beyond 30 m.

On the basis of the laboratory data, some improvement in performance can be expected if the strips are distributed as patches, for example, a chequerboard-type of arrangement, because this would mean that the dependence of the reduction on the azimuthal angle would be decreased.

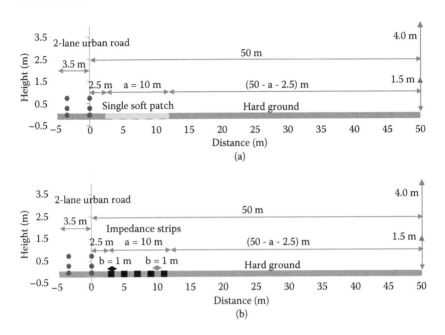

Figure 7.9 Schematic diagrams (a) showing a continuous strip of "soft" ground (such as gravel) next to a two-lane urban road; (b) an example distribution of alternating equally wide strips of "soft" and "hard" ground (*a* = 10 m and *b* = 1 m). Receivers are located at 50 m distance from the edge of the road and are 1.5 m and 4 m high.

7.2 PREDICTED EFFECTS OF GROUND TREATMENTS AROUND RAILWAYS

7.2.1 Introduction of grassland

Figure 7.10 includes a vertical cross section through a railway with four rail/wheel contact sources, acoustically hard sleepers, and profiled ballast. Each source is assumed to be at a height of 5 cm above the top of 0.171-m-high rails. Note that the acoustically soft ground is assumed to start 5.0 m from the nearest edge of the ballast instead of at 2.5 m, as for other abatement types, and to extend to the 1.5-m-high receiver either 49.5 m or 102.5 m from the nearest edge of the ballast profile. Four soft ground types (sites 41, 16, 22, and 24) have been chosen for calculations with acoustical properties described by the two-parameter slit pore impedance model parameters specified in Table 7.2. The properties assumed for the ballast are listed in Table 7.10. Figure 7.11 shows two rail noise spectra: one recorded in the United Kingdom 1 m away from a train moving at 145 km/h[11] and the other used as a reference railway noise spectrum in HOSANNA.[12]

Table 7.9 Predictions of the insertion loss due to a single 9-m-wide strip of "gravel" or "grass" (i.e., b = 0 m) with x = 2.5 m and nine alternating soft/hard strips of width 1 m alongside a two-lane urban *road*

Strip widths and configuration		Receiver height (m)	Insertion loss (dB)					
			Semi-infinite "Gravel" (Flow resistivity = 10 kPa s m^{-2}, Porosity = 0.4)			Semi-infinite "Grass" (Flow resistivity = 150 kPa s m^{-2}, Porosity = 0.5)		
a (m)	b (m)		Lane 1	Lane 2	Combined	Lane 1	Lane 2	Combined
5.0	0	1.5	4.5	3.9	4.2	3.6	3.1	3.4
		4.0	4.1	2.7	3.4	3.4	2.4	2.9
10.0	1.0	1.5	4.8	4.2	4.5	3.6	3.1	3.4
		4.0	3.8	2.5	3.1	3.1	2.2	2.7
9.0	0	1.5	6.4	5.6	6.0	5.1	4.6	4.9
		4.0	4.4	2.8	3.5	3.9	2.7	3.3
15.0	0	1.5	8.2	7.3	7.7	6.8	6.1	6.5
		4.0	4.0	2.5	3.2	3.9	2.6	3.2
25.0	1.0	1.5	8.0	7.1	7.5	6.1	5.6	5.8
		4.0	3.6	2.2	2.8	3.5	2.3	2.9
25.0	0	1.5	9.5	8.6	9.1	8.1	7.5	7.8
		4.0	3.8	2.3	3.0	4.0	2.6	3.3

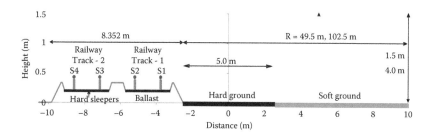

Figure 7.10 A railway track profile showing four rail/wheel sources, profiled ballast, and soft ground extending from 5 m up to 102.5 m from the foot of the ballast to the 1.5-m- or 4-m-high receiver.

Table 7.10 Parameter values used for calculating the acoustical properties of railway ballast and porous concrete

Material	Flow resistivity (kPa s m^{-2})	Porosity	Tortuosity	Viscous characteristic length (m)
Railway ballast	0.2	0.491	1.3	0.01
Porous concrete	3.619	0.3	1.8	2.2 × 10^{-4}

Source: Attenborough, K., P. Boulanger, Q. Qin, and R. Jones. 2005. Predicted influence of ballast and porous concrete on rail noise. Presented at the *Inter-Noise 2005*, I-INCE, August 7–10, Rio de Janeiro. With permission.

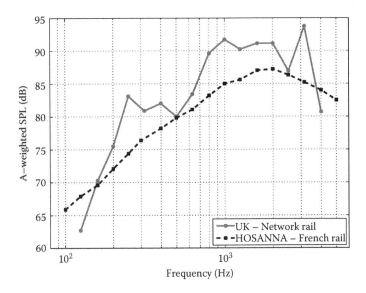

Figure 7.11 Railway noise spectra at 1 m assumed for the calculations reported in Table 7.10.

Table 7.11 shows results of predictions using BEM. At a 1.5-m-high receiver, the predicted insertion losses vary between about 3 dB at 50 m from the nearest track and 11 dB at 100 m from the farthest track. The corresponding predicted insertion loss values at a 4-m-high receiver are between less than a dB and a little over 3 dB. The predicted insertion losses using the U.K. train noise spectrum are approximately 2 dB higher than those obtained when using the HOSANNA train noise spectrum.

As predicted for road traffic noise, introduction of lower flow resistivity grassland results in higher insertion losses at the 1.5-m-high receiver. For the closest track, there is between a 3- and 4-dB difference between lowest and highest predictions of insertion loss.

7.2.2 Gravel strips

Predictions also have been made for the insertion loss resulting from the introduction of continuous gravel (flow resistivity 10 kPa s m^{-2}, porosity 0.4, total width a m) or parallel gravel strips of width b m, starting 2.5 m from the nearest edge of the ballast profile (see Figure 7.10). For the U.K. spectrum, the closest track and the 1.5-m-high receiver at 102.5 m from the nearest edge of the ballast, the predicted insertion losses are between 3 dB for a 5-m-wide gravel area and 9 dB for a 25-m-wide gravel area. The predictions using the HOSANNA rail noise spectrum

Table 7.11 Predicted insertion loss due to soft ground near railway tracks

Soft ground Type	Range "R" (m)	Receiver height "Hr" (m)	Insertion loss (dB)—Frequency range: 25–10 kHz U.K. spectrum						Insertion loss (dB)—Frequency range: 25–10 kHz HOSANNA spectrum					
			Source -1	Source -2	Combined 1 & 2	Source -3	Source -4	Combined 3 & 4	Source -1	Source -2	Combined 1 & 2	Source -3	Source -4	Combined 3 & 4
Long grass (NT site #41)	49.5	1.5	7.1	7.2	7.1	5.0	5.4	5.2	5.0	4.7	4.9	3.2	3.1	3.1
		4.0	1.1	1.0	1.0	0.4	0.4	0.4	1.2	0.9	1.1	0.3	0.2	0.2
	102.5	1.5	12.9	12.7	12.8	10.4	10.7	10.6	9.6	9.1	9.4	7.4	7.2	7.3
		4.0	5.1	5.4	5.2	3.0	3.6	3.3	3.8	3.4	3.7	1.9	2.0	1.9
Lawn (NT site #16)	49.5	1.5	7.2	7.2	7.2	5.0	5.4	5.2	5.0	4.6	4.9	3.1	3.0	3.1
		4.0	1.1	1.1	1.1	0.4	0.3	0.4	1.2	1.0	1.1	0.2	0.2	0.2
	102.5	1.5	13.0	12.7	12.9	10.3	10.5	10.4	9.6	9.1	9.4	7.3	7.1	7.2
		4.0	5.3	5.6	5.4	3.1	3.6	3.3	3.8	3.4	3.7	1.9	1.9	1.9
Pasture (NT site #22)	49.5	1.5	5.8	5.1	5.5	3.7	3.7	3.7	4.3	3.6	4.0	2.5	2.3	2.4
		4.0	1.1	1.1	1.1	0.3	0.3	0.3	1.2	0.9	1.1	0.2	0.2	0.2
	102.5	1.5	9.5	8.0	8.9	6.8	6.3	6.6	8.0	7.0	7.6	5.6	5.3	5.5
		4.0	4.9	4.6	4.8	2.7	2.9	2.8	3.7	3.0	3.4	1.8	1.6	1.7
Arable (NT site #24)	49.5	1.5	5.1	4.3	4.8	3.3	3.2	3.2	3.9	3.2	3.6	2.3	2.0	2.2
		4.0	1.0	1.0	1.0	0.3	0.3	0.3	1.1	0.9	1.0	0.2	0.2	0.2
	102.5	1.5	8.3	6.9	7.8	6.0	5.5	5.8	7.3	6.2	6.9	5.2	4.8	5.0
		4.0	4.5	4.1	4.4	2.5	2.7	2.6	3.5	2.8	3.2	1.7	1.5	1.6

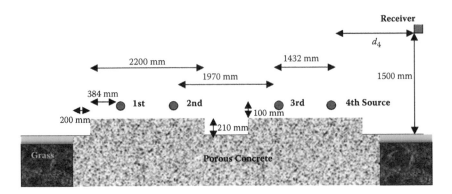

Figure 7.12 Slab track profile and assumed source/receiver geometry.

are between 1 and 2 dB lower. As is the case with road traffic noise, there is no predicted advantage from introducing strips of gravel rather than continuous gravel.

7.2.3 Porous concrete slab track

Figure 7.12 shows the track profile and source–receiver geometry assumed for predicting the effect of replacing an acoustically hard slab track by a porous concrete track in a railway for high-speed trains. Figure 7.13(a)(b) compares predicted spectra for acoustically hard and acoustically soft slab track using the U.K. and HOSANNA source spectra, respectively. The assumed porous concrete properties are listed in Table 7.10.

The insertion loss predicted using the U.K. spectrum, which has low and high frequency peaks, is higher. For a train on the nearest track, the insertion loss is 5.5 dB and for a train on the farthest track, the insertion loss is nearly 3.0 dB.

7.3 ROAD TRAFFIC NOISE REDUCTION USING BURIED RESONATORS

A resonator consists of a hollow container connected to its surroundings through a neck (rather like an empty bottle). The resonance frequency, sometimes called the natural frequency, depends on the cross-sectional area of the neck and the neck length as well as the volume of the container. Often, sound-absorbing material is placed inside the resonator. However, even without such material, the resonator absorbs sound energy at and around the natural frequency by converting it to heat through viscous and thermal processes in the resonator neck. As sound passes over a 'bottle-type' resonator, the 'plug' of air in the resonator

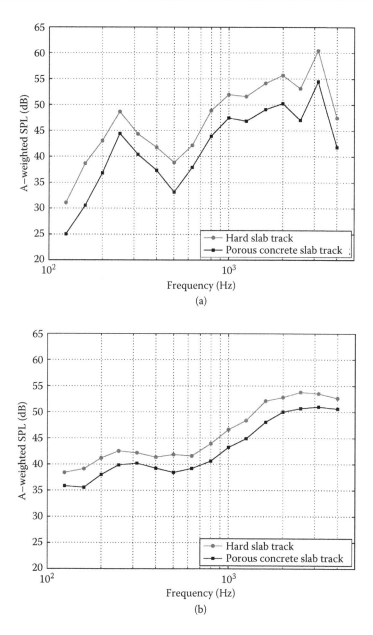

Figure 7.13 Comparison between BEM-predicted A-weighted sound pressure level (SPL) spectra at 25 m from the nearest track over acoustically hard and porous concrete slab tracks using the profile shown in Figure 7.11, and porous concrete properties listed in Table 7.9. (a) using U.K. source spectrum; (b) using HOSANNA source spectrum-combined effects of sources at positions 1 and 2.

neck moves up and down repeatedly and experiences resistance due to viscous drag in the neck and due to the compressibility of the air in the main body of the resonator, which is 'stiff' as a result of being more or less enclosed (like in a balloon). The resonance frequency depends on the ratio of the stiffness of the air inside the resonator to its total volume. An array of buried resonators in an otherwise acoustically hard area can cause noise reduction.

7.3.1 Resonators buried in porous road surfaces

It is possible to combine acoustic resonators with porous road surfaces (Figure 7.14). Buried resonators affect the acoustical properties of a porous asphalt road in two ways: (1) they increase absorption during propagation over the road surface, and (2) they reduce the sound amplification associated with the geometry of the tyre–road contact (the "horn effect").

The sound absorption coefficient measured perpendicular to the surface of twin-layer porous asphalt with an overall layer thickness of approximately 7 cm has pronounced maxima at about 600 Hz and 1,800 Hz. These maxima decrease with decreasing angle of incidence. Nevertheless, the frequency dependence of the absorption coefficient can be improved by inserting resonators tuned to 1 kHz (as shown by calculations and measurements with and without resonators in Figure 7.15(a)). Buried resonators can improve the noise reduction of new twin-layer porous asphalt by approximately 3 dB.

Pass-by measurement results according to ISO 11819-1[13] on a test section of a highway containing buried resonators after three years under traffic use have shown that the buried resonators still lead to a sound pressure level reduction (compared to the situation without resonators) of about 3–4 dB for passenger cars (Figure 7.15(b)) and about 2 dB for heavy trucks for a 1.2-m-high receiver at 7.5 m distance.

Figure 7.14 Example of a resonator covered with porous asphalt (a), and the construction of a resonator-improved porous asphalt (b).

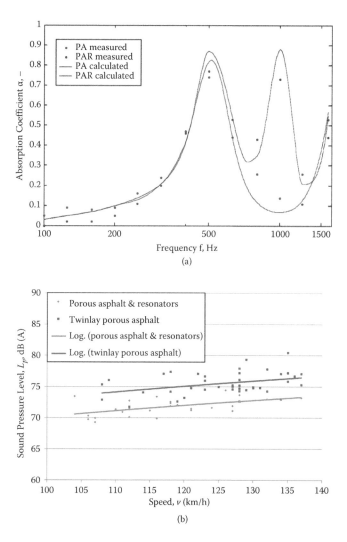

Figure 7.15 (a) Absorption coefficient of porous asphalt at gracing sound incidence with and without resonator improvement compared to a normalised traffic noise spectrum. (b) Measurement results of pass-by measurements on porous asphalt with and without resonators.

7.3.2 Resonators buried in hard ground

Burying an array of resonators in an otherwise acoustically hard surface offers another means for reducing traffic noise. This idea has been tested in the laboratory. A surface-treated plywood plate, containing a 7 × 7 square array of 6 mm diameter holes at 30 mm centre-to-centre spacing, was placed over a 30-mm-deep square array of cavities formed with aluminium walls, which, in turn, rested on a large (1.2 m × 1.1 m) plywood plate. Measurements of the sound pressure spectra due to a point source have been made with and without the holes. The measurements without holes were made after covering them with tape. The point source, consisting of a compression driver connected to an approximately 3-m-long flexible hose with an aluminium ring at the opening and an inner diameter of 25 mm, was located at a height of 60 mm above the plate containing holes and 600 mm from its centre. The receiver was placed on the plate containing holes, 30 mm behind the last row of holes, and on axis so that the source–receiver line coincided with the plate centre line. Figure 7.16 shows parts of the measurement setup.

Figure 7.17(a)(b) shows measured excess attenuation spectra and predictions obtained from a lumped element resonator theory for each resonator combined with the numerical solution for an array of resonators. Both measurements and predictions show that the (positive) insertion loss increases as the hole (neck) diameter is increased. Also, the larger the neck diameter, the higher is the resonance frequency.

Calculations have been made of the noise reduction due a 4-m-wide strip containing a square array of resonators with centre-to-centre spacing of 6 cm and neck opening of 2 cm tuned to 380 Hz, starting 2.5 m

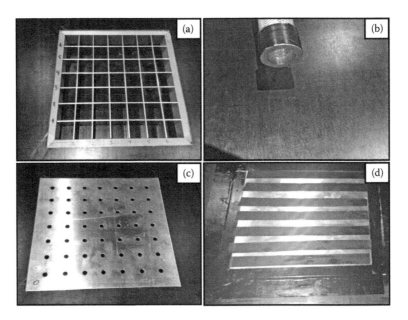

Figure 7.16 Parts of the laboratory array of resonators: (a) backing cavities, (b) source, (c) top plate with hole array, and (d) top plate with holes covered by tape.

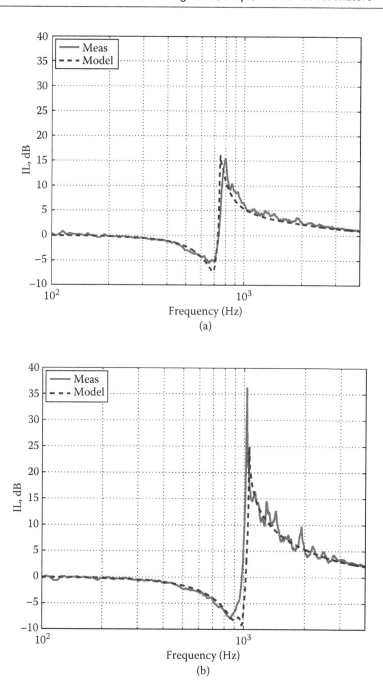

Figure 7.17 Excess attenuation due to a laboratory resonator array: (a) hole diameter 6 mm, and (b) hole diameter 10 mm.

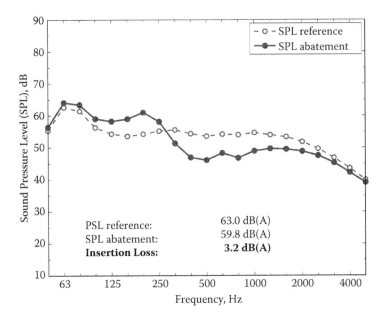

Figure 7.18 Calculated sound pressure levels (SPL) at a 1.5-m-high receiver due to a two-lane urban road, with and without resonators, at a 50-m range.

from the centre of the nearest lane of a two-lane urban road (95% cars type 1 and 5% heavy vehicles travelling at a mean speed of 50 km/h). At a 1.5-m-high receiver 40 m away, the insertion loss is predicted to be nearly 3 dB (Figure 7.18).

7.4 CONCLUSION

Ground effects can be exploited for noise control. Even though it is difficult to achieve as much reduction as might be obtained with a traditional noise barrier of 2-m height, the advantage of exploiting ground effect is that it does not create an impassable division between communities. Replacing acoustically hard ground by acoustically soft ground offers the opportunity of adding to the "green" in cities, which can be augmented in an acoustically beneficial way by crops or other dense vegetation up to 1 m in height. Careful choice of soft ground can make a significant difference to the resulting noise reduction. Porous road surfaces are used already as a means of road traffic noise reduction, but their performance can be improved by incorporating preformed resonating chambers. Buried resonators can be used to increase attenuation over nonporous, acoustically hard ground.

REFERENCES

1. Attenborough, K., I. Bashir, T. J. Hill, S. Taherzadeh, J. Defrance, and P. Jean. 2014. Noise reduction using surface roughness. In *Environmental methods for transport noise reduction*, eds. M. E. Nilsson, R. Klæboe, and J. Bengtsson (pp. 122–154). Oxford, U.K.: Spon Press.

2. ISO. 1996. *Acoustics—Attenuation of sound during propagation outdoors—Part 2: A general method of calculation (ISO 9613-2)*. Geneva: International Standards Organization.

3. Attenborough, K., K. M. Li, and K. Horoshenkov. 2007. *Predicting outdoor sound*. London: Taylor & Francis.

4. Nota, R., R. Barelds, and D. Van Maercke. 2005. Harmonoise WP 3 engineering method for road traffic and railway noise after validation and fine tuning. Deliverable of WP3 of the HARMONOISE project. Document ID HAR32TR-040922-DGMR20.

5. Attenborough, K., I. Bashir, and S. Taherzadeh. 2011. Outdoor ground impedance models. *The Journal of the Acoustical Society of America* 129(5): 2806–2819.

6. de Jong, B. A., A. Moerkerken, and J. D. van der Toorn. 1983. Propagation of sound over grassland and over an earth barrier. *Journal of Sound and Vibration* 86(1): 23–46.

7. Van Renterghem, T., K. Attenborough, and P. Jean. 2014. Designing vegetation and tree belts along roads. In *Environmental methods for transport noise reduction*, eds. M. E. Nilsson, R. Klæboe, and J. Bengtsson (pp. 93–118). Oxford, U.K.: Spon Press.

8. Van Maercke, D., J. Defrance, J. Maillard, C. Anselme, A. Mandon, B. Altreuther, Y. Smyrnova, T. Van Renterghem, J. Y. Jeon, H. S. Jang, and Y. H. Kim. 2013. *Technical report for task 6.1 (FP7 HOSANNA Deliverable 6.1)*. Online at: http://www.greener-cities.eu/

9. Aylor, D. 1972. Sound transmission through vegetation in relation to leaf area density, leaf width, and breadth of canopy. *The Journal of the Acoustical Society of America* 51 (1B): 411.

10. Aylor, D. 1972. Noise reduction by vegetation and ground. *The Journal of the Acoustical Society of America* 51(1B): 197.

11. Attenborough, K., P. Boulanger, Q. Qin, and R. Jones. 2005. Predicted influence of ballast and porous concrete on rail noise. Presented at *Inter-Noise 2005*, I-INCE, August 7–10, Rio de Janeiro.

12. Defrance, J., P. Jean, F. Koussa, K. Horoshenkov, A. Khan, H. Benkreira, T. Van Renterghem, J. Kang, Y. Smyrnova, and J. Forssén. 2013. *Application to innovations (FP7 HOSANNA Deliverable 2.3)*. Online at: http://www.greener-cities.eu/

13. ISO. 1997. *Acoustics—Measurement of the influence of road surfaces on traffic noise—Part 1: Statistical pass-by method (ISO 11819-1)*. Geneva: International Standards Organization.

Chapter 8

Vegetation in urban streets, squares, and courtyards

Jian Kang, Maarten Hornikx, Timothy Van Renterghem,
Yuliya Smyrnova, Jens Forssén, Chris Cheal,
Dick Botteldooren, Hong-Seok Yang, Jin Yong Jeon,
Hyung Suk Jang, Shahram Taherzadeh,
Keith Attenborough, and Agnes Mandon

Abstract. One of various ways in which vegetation cover used in the greening of urban areas can help improve the health and well-being of people is in how it changes the acoustic environment. This chapter presents findings of computer simulations and scale modelling to examine and quantify the effectiveness of green roof and green wall (vertical garden) systems in reducing road traffic noise for streets, squares, and roadside courtyards. Noise reduction by sound absorption in reflected and diffracted (over roofs) sound paths is investigated. Particular attention is paid to the importance of vegetation placement relative to the receiver/listening positions. Because the soil substrate used for the vertical walls has good sound absorption properties, it also can be used for green barriers. In this chapter, the effects of a low barrier made of green wall substrate are studied for an installation on the ground and on the top of buildings surrounding a courtyard.

8.1 ACOUSTIC POTENTIAL OF GREEN ROOF AND GREEN WALL SYSTEMS IN THE URBAN CONTEXT

This chapter focuses on the expected reductions in traffic noise in urban streets, courtyards, and squares after installing green roof and green wall systems to building envelopes. In the context of urban streets, belts of trees, bushes, and seasonal flowers are the common types of vegetation seen on roadsides. Urban squares may feature patches of grass, individual trees, shrubs, or small parks. In densely built urban centres, building envelopes are often the only places offering opportunities to increase the amount of greening essential for recovering ecological balance and providing visually pleasant environments. Ecological advantages include increasing the thermal insulation of the building envelope,[1] reduction of the urban heat island effect,[2] acting as a buffer for stormwater,[3] improving air quality,[4] and providing habitat space for birds[5] and beneficial insects.[6] The potential of a green roof and a green wall system to reduce noise also has been recently identified.[4,7,8]

The green roof and green wall systems investigated include the Zinco green roof and the Canevaflor green wall substrates, both 20 cm in depth, whose sound absorption properties have been studied in laboratories of the University of Bradford (West Yorkshire, U.K.) and University of Sheffield (Sheffield, U.K.), respectively, in accordance to ISO 10534-2[9] and ISO 354:2003[10] (cf. Van Renterghem et al.,[11] and in Chapter 4 of this volume). The absorption coefficient of green roof substrate is small, between 0.08–0.2, in the low frequency range (i.e., below 125 Hz). Thereafter, it gradually increases with frequency and is about 0.6–0.7 at frequencies above 200 Hz. In the case of green wall substrate, the absorption coefficient increases from 0.1 to 0.8 in the frequency range of 50–100 Hz, it then drops to 0.6 at 200 Hz, and reaches a stable value of about 0.7 at frequencies above 200 Hz. All data described above are representative for substrates that are dry. Adding moisture to them had a small effect on the absorption coefficient, which is true for the Canevaflor green wall substrate, but not true for green roofs, although other types of substrate may exhibit higher dependence on the water content. This characteristic should be considered when selecting substrate partly as a noise abatement measure in an urban environment. Also, the coverage of substrates by vegetation brings a slight reduction in absorption coefficient at higher frequencies, but some improvements at lower frequencies.[12]

Acoustic effects of vegetation in typical urban spaces, such as streets, squares, and courtyards, are related to three mechanisms at the boundaries between different media:[13] (1) sound absorption; (2) sound diffusion, which occurs when a sound wave impinges on the vegetation and is then reflected; and (3) sound transmission when a sound wave is passing through the vegetation. Increasing boundary absorption in urban spaces can reduce the sound pressure level—measured in decibels (dB) either unweighted or weighted with the A-filter, dB(A), that corrects for the frequency-dependent sensitivity of the human hearing system. Application of green wall and green roof systems on building envelopes can have such an effect, and this could be greatly enhanced in urban areas because there are many boundaries producing multiple reflections. Compared with boundaries reflecting sound in one direction (specularly), boundaries reflecting sound diffusely in many directions, such as around vegetation, may decrease reverberation time (i.e., the time required, in seconds, for the average sound in a space to decrease by 60 decibels after a source stops generating sound) and increase sound attenuation especially lengthways in a street canyon.[14] In urban areas, due to the presence of multiple reflections, the diffusion effect of vegetation will be greater, even when the diffuse part of a surface reflection is relatively small. In the case of sound from a nearby trafficked street canyon being diffracted by buildings surrounding a courtyard and bending over the buildings, thus entering the courtyard, a green roof on the building may absorb and scatter the diffracted sound and decrease noise in the courtyard.

8.2 STUDIED CASES

To demonstrate the effects of green roof and green wall systems in reducing urban traffic noise, typical cases from European cities were selected for consideration as reference configurations *without* vegetation. Section 8.2.6 describes how these reference configurations are modified by the introduction of green measures and reexamined as configurations *with* vegetation.

8.2.1 Reference configurations

In each of the following reference configuration cases (A–E), a 19.2-m-high, six-storey building is assumed with all façades consisting of brickwork elements and windows. The windows, equally distributed on the building façades, are 1.9 m high and have an outside depth (reveal) of 0.16 m to introduce diffusivity that is nearly always present in the sound field in urban spaces. The windows have been assigned a frequency-independent absorption coefficient of 0.1. A rather high frequency-independent absorption coefficient for the façades of 0.33 has been assumed following recommendations in ISO 9613-2.[15] This value includes, to some degree, small-scale sound scattering at common building materials, which can be mimicked by increasing the absorption coefficients.

For the scale modelling (A1, C1, C7), the reference configuration is assumed with a 10.0-m-high, three-storey building with the same façade shape as that in the simulation cases, and the width of the road is 9.2 m, with two lanes. Acoustically, rather hard surfaces on the façades are assumed with an absorption coefficient of 0.1.

8.2.2 Case A: single street

The street is 76.8 m long (60.0 m long in the scale model) and 19.2 m wide (9.2 m wide in the scale model) and has two parallel opposing traffic lanes separated by a 4.0-m-wide paved central reservation (median). By assuming periodic boundary conditions in the y-direction (Figure 8.1), a longer street is actually modelled. Receiver points are located along two facades and at a horizontal line at a height of 1.5 m.

8.2.3 Case B: urban square with a trafficked street on one side

Alongside streets, squares represent another typical morphology of the urban layout. In many European cities, traffic runs through a street adjacent to the square as it is considered in Figure 8.2(a)(b).

The square is 38.4 m wide (x-direction) and 57.6 m long (y-direction) and the street is 19.2 m wide and 115.5 m long. Four traffic lanes shown

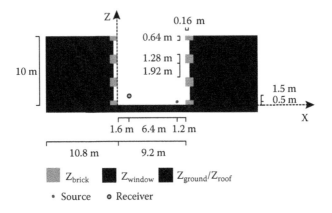

Figure 8.1 Cross section of Case A.

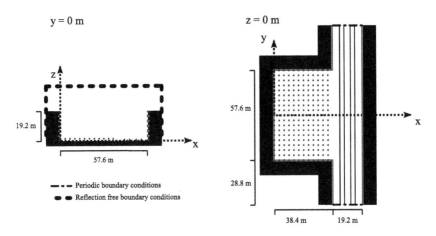

Figure 8.2 Cross section (a) and plain view (b) of Case B. Red lines represent traffic lines, green dots represent receiver positions. Periodic boundary conditions are imposed in *y*-direction.

in Figure 8.2(b) in red run along the street. The receivers are located in a horizontal plane at a height of 1.5 m.

8.2.4 Case C: street with a completely enclosed courtyard

A street with a crossing (intersecting) minor street and a completely enclosed roadside courtyard is shown in Figure 8.3. The trafficked street is 19.2 m wide and 52.8 m long, while the crossing street is 9.6 m wide

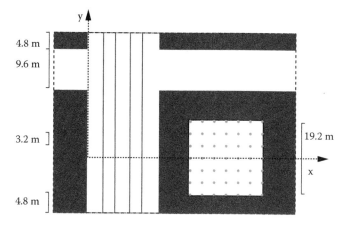

Figure 8.3 Plan view of Case C. Red lines represents traffic lanes, green dots represent receiver positions. Periodic boundary conditions are imposed in y-direction.

and 67.2 m long. The square courtyard has sides of 19.2 m. Receivers are equally distributed in the courtyard at a height of 1.5 m. Receivers along a vertical line on the courtyard façades also are considered in the simulations.

8.2.5 Case D and Case E: street and a courtyard with a façade opening

Roadside courtyards are not always completely closed and façade openings for convenient accessibility may be encountered. By transmitting too much noise from road traffic, such façade openings may prohibit the use of these courtyards as quiet areas.

A street with a crossing street and a courtyard with a 3-m-wide opening is considered (Figure 8.4(a)(b)). In Case D, the opening is facing the cross street, while in Case E it is facing the trafficked street. Two heights of the opening, 3 m and 19.2 m (equal to the building height), are considered.

8.2.6 Configurations with green measures

The acoustic properties of the green roof and green wall substrates selected as green measures to abate noise in the reference configurations were described in Section 8.1. Green roofs were applied directly to building roofs, while green walls were applied to building façades and all sides of low-height screens were placed on the street and the roof of the buildings.

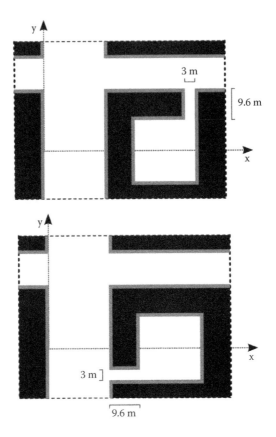

Figure 8.4 Plan view of Case D (top) and Case E (bottom).

Figure 8.5 presents an overview of these measures. In the case of 3D calculations, the roof edge screens and green roofs are applied to the full roof surface.

In Cases A and B, the green wall substrate has been mounted on the brickwork parts of the façades. Three positions of the vegetated wall have been studied: on the whole façades, on the upper halves of the façades, and on the low parts of the façades. In Cases A4 and B4, a low-profile vegetated barrier of dimensions 1 m × 1 m (width × height) made from green wall substrate is considered. In Case A4 the barrier runs along the entire length of the street in the central reservation between two lanes of traffic, while, in Case B4, a vegetated barrier is placed between the square and the adjacent street.

In configurations C1–C15, single measures with either green roof or green wall are assumed, while in C16–C19, combined measures are considered.

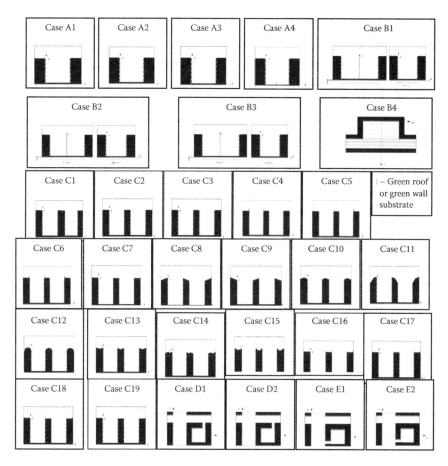

Figure 8.5 Vegetation measures evaluated in reference Cases A–E. Al, Cl, C7 are evaluated using scale models with lower street canyon and courtyard.

8.3 TRAFFIC MODEL AND PREDICTION APPROACHES

The following simulations of traffic noise abatement by vegetation utilise sound source models based on the characteristics of road traffic in European urban configurations. These models are based on data from a systematic study of the type, speed, and presented in the HARMONOISE research project report.[16] This section provides an overview of the simulation approaches applied to predict the effectiveness of the green measures to reduce traffic noise in the selected case studies. Measures that have been calculated to predict the effect of noise mitigation are also presented.

8.3.1 Traffic model

A road traffic lane has been modelled by three sound source heights above the ground at 0.01 m, 0.30 m, and 0.75 m. Traffic is composed of 95% light vehicles and 5% heavy vehicles and with speeds of 30 km/h, 40 km/h, 50 km/h, 60 km/h, and 70 km/h. For the scale model, ribbon tweeters with flat frequency response at the high frequency bands for scale modelling were used as a line source to simulate road traffic noise. The height of sound source is 0.5 m above the ground.

8.3.2 Prediction models

Sound propagation in urban spaces is complex, usually including a combination of multiple reflections, diffraction, scattering, and other physical effects. Accurate predictions require the application of models capable of simulating these effects in an integrated way. Wave-based numerical models, therefore, are needed. FDTD (finite-difference time domain) and PSTD (pseudo-spectral time domain) were combined. Both solve the same physical sound propagation equations.[17–19] PSTD allows full 3D calculations, while with ordinary computing power, FDTD is usually limited to 2D applications. On the other hand, FDTD allows a more advanced treatment of material boundaries.[7,20,21] For the purpose of studies presented in this chapter, computations with the FDTD and PSTD methods are performed between the frequencies 50 Hz and 1.6 kHz, which can be considered to be sufficient at highly shielded locations in an urban environment.

Although wave-based modelling approaches are needed to predict sound propagation at low frequencies, they are less applicable to simulate phenomena at higher frequencies due to high computation demands. Models similar to the geometrical optics approach are more applicable for these frequencies. One such model is the CRR (combined ray-tracing and radiosity) model. In CRR, the *ray-tracing model* is applied to simulate specular reflections, whereas the *radiosity model* simulates scatter reflections from the space boundaries.[22] CRR model/software was developed at the University of Sheffield and has been verified in simulations of various urban spaces.[14,22] Computations with CRR methods in this chapter are performed between 100 Hz and 4 kHz.

Scale modelling has been an effective tool for evaluating the complex behaviours of sound, such as reflection, scattering, and diffraction in street canyons and courtyards.[23–25] This technique also is useful in considering complex vegetation shapes with leaves and stems. In this study, a 1:10 scale model was constructed and the frequency range considered was 100–4 kHz (full scale).

8.3.3 Measures of green roof and green wall effects

The results of noise mitigation by the introduction of vegetation are studied in terms of *insertion loss* $(\overline{IL})$ in dB(A) obtained by calculating the difference between A-weighted sound pressure levels in a reference configuration and a configuration with vegetation averaged over all receiver positions and expressed as:

$$\overline{IL} = \frac{1}{J}\sum_{j=1}^{J}\left(L_j^{ref} - L_j\right), dB(A),\qquad(8.1)$$

where J is the number of receiver positions and *ref* refers to the reference configuration, and L_j^{ref} and L_j are road traffic sound pressure levels obtained at *j*th receiver position in, respectively, the reference configuration and a configuration with vegetation.

Similarly, insertion loss values also are obtained in each frequency band of interest and are measured in dB:

$$\overline{IL_k} = \frac{1}{J}\sum_{j=1}^{J}\left(L_{j,k}^{ref} - L_{j,k}\right), dB,\qquad(8.2)$$

where $L_{j,k}^{ref}$ and $L_{j,k}$ are sound pressure levels obtained at *j*th receiver position for the kth 1/3 octave band.

8.4 EFFECT OF VEGETATION

This section examines and quantifies the effectiveness of vegetation in reducing traffic noise in urban streets, squares, and courtyards, while considering the effect of absorption and scattering. A parametric study is undertaken that demonstrates, examines, and quantifies the effectiveness of vegetation in noise reduction, as well as the effectiveness of strategic design and arrangements of vegetation within different street/square dimensions and noise abatement schemes using vegetation.

In Table 8.1, insertion losses are presented for all configurations with green measures considered in Section 8.2.6 where green roof and green wall are denoted as GR and GW, respectively. Insertion losses obtained for 2D, 3D modelling approaches by simulation and scale modelling (SC) are indicated accordingly.

8.4.1 Case A

The simulation results suggest that the noise reduction is below 1 dB due to dominate effect of direct sound and relatively low impact of reflected sound for the line source in the street. Greening the façades of the street mostly

Table 8.1 Overview of insertion losses $\overline{IL}$ obtained for different green measures: green roof (GR), screen (S), and green wall (GW)

Case ID	Treatment type	$\overline{IL}$, dB(A)	Case ID	Treatment type	$\overline{IL}$, dB(A)
A1	GW	1.0 (3D)	C9	GR	6.0 (2D)
		1.6 (SC)			7.0 (3D)
A2	GW	0.6(3D)	C10	GR	7.7 (2D)
					8.2 (3D)
A3	GW	0.3 (3D)	C11	GR	5.0 (2D)
					5.6 (3D)
A4	GW	2.0 (3D)	C12	GR	6.3 (2D)
					6.5 (3D)
B1	GW	1.0 (3D)	C13	GR	2.1 (2D)
					2.7 (3D)
B2	GW	0.8 (3D)	C14	GR	3.0 (2D)
					3.7 (3D)
B3	GW	0.55 (3D)	C15	GR	0.6 (2D)
					0.8 (3D)
B4	GW	4.5 (3D)	C16	S + GR	4.4 (2D)
					4.4 (3D)
C1	GW	1.6 (2D)	C17	S + GW	4.1 (2D)
		2.0 (3D)			3.3 (3D)
		2.2 (SC)			
C2	GW	1.1 (2D)	C18	S + GR + GW	5.1 (2D)
		1.3 (3D)			4.6 (3D)
C3	GW	0.6 (2D)	C19	GR + GW	2.1 (2D)
		0.5 (3D)			2.2 (3D)
C4	S	1.5 (2D)	D1	GW	3.4 (3D)
		0.8 (3D)			
C5	S	1.6 (2D)	D2	GW	4.1 (3D)
		0.9 (3D)			
C6	S	3.6 (2D)	E1	GW	4.5 (3D)
		2.8 (3D)			
C7	GR	2.3 (2D)	E2	GW	4.1 (3D)
		2.7(3D)			
		1.7 (SC)			
C8	GR	6.2 (2D)			
		7.4 (3D)			

affects the highest frequencies. In Case A1, the highest $\overline{IL_k}$ is of 1.2–1.3 dB for the middle and high frequency ranges (400–4 kHz). If only the lower halves of the façades are covered with vegetation (A2), the reduction is about 0.6 dB(A), whereas vegetation on the upper halves (A3) of the façades may result in a reduction of about 0.3 dB(A).

Simulation results are dependent to a large degree on the acoustic properties assumed for underlying materials in the reference configurations (i.e., those without vegetation). Acoustically hard surfaces, like concrete or glass, reflect more sound than softer surfaces, such as wood or vegetation. In these studies, a relatively acoustically soft brickwork surface with absorption coefficient of 0.33 was simulated as reference. Decreasing absorption coefficient to 0.1 leads to an increase in the sound pressure level (SPL) in the reference configuration and, as a consequence, produces a greater difference between it and the SPL in the configuration with vegetation (i.e., a larger insertion loss) with an average increase in IL of about 2.0 dB(A).

Inserting a low-profile vegetated screen in the street (A4) gives an average noise reduction of about 2 dB(A). The largest noise reduction ($\overline{IL_k}$) is obtained for lower receiver position (at 1.5 m above the ground), 7 dB for 500 Hz for example which is mainly due to the shielding effect and greater absorption of the green wall at higher frequency range compared to frequencies <100 Hz.

8.4.2 Case B

A reduction of traffic noise of 1.0 dB(A) averaged over all receivers is achieved with green wall covering all façades in the square and the adjoining street (B1). Placing a green wall on lower (B2) and upper (B3) parts of the façades in the square and street gives insertion losses of 0.8 dB(A) and 0.5 dB(A), respectively.

Assuming an absorption coefficient of brickwork in the reference configuration of 0.1 results in an average increase in IL of 2.5 dB(A) for the square case (B1).

In the square, for the case with all façades treated with vegetation (B1), the largest insertion loss (IL) is obtained for the receiver positions located far from the traffic and close to the façades with vegetation (2.0 dB at 500 Hz and 3.5 dB at 1 kHz).

As in the case of urban streets, the noise reduction performance of vegetated façades is greater for narrower squares. Note that if traffic runs through the square itself, the vegetation will reduce the noise to a smaller extent by less than in the case of the traffic running on a side street.

Insertion of a vegetated barrier between the square and the adjacent street (B4) may reduce noise by up to 4.5 dB(A). Depending on frequency, an increasing noise reduction in the range of 2–8 dB can be obtained in the frequency range above 100 Hz for the receiver positions close to the barrier and this indicates the shielding effect.

8.4.3 Case C: vegetated courtyard façades

Having a quiet side bordering a dwelling would be useful in reducing the adverse effects of noise, such as annoyance and sleep disturbance.[26-28] Methods that reduce noise in courtyards, therefore, can be valuable as a complement to noise reduction on the most noise-exposed façades of buildings. Vegetated façades in courtyards reduce noise from all sources situated outside the courtyard, and they are beneficial as well for noises originating from inside the courtyard.

Completely greening the courtyard façades (C1) gives an average noise reduction of 2 dB(A). Applying a green wall on only the upper half (C2) results in average insertion loss of 1 dB(A), while greening the lower half (C3) gives about 0.5 dB(A) of noise reduction.

The effect of vegetated courtyard façades is greatest at the highest frequencies and for lower receiver positions. In an elongated courtyard next to a trafficked street, with the longer side exposed to the traffic, vegetated façades reduce the noise only slightly less than for case C, with a difference below 0.5 dB(A). Higher noise reduction is expected for streets with half the width and courtyards with half the area, but the additional reduction is under 1 dB(A).

Changing the underlying material in the reference configuration to a harder one with absorption coefficient of 0.1 additionally gives an average increase in noise reduction by about 2 dB(A).

8.4.4 Case C: vegetated roof barriers

A vegetated roof barrier improves the sound field in the courtyard by increasing absorption near the diffraction edges. For the simulation, a barrier of width 0.6 m and height of 1 m is chosen and all barrier surfaces are treated with the green wall substrate. As Table 8.1 shows, the use of a single roof screen near the source street canyon edges (C4), or near the receiver street canyon edges (C5), only gives limited and very similar noise reduction of about 1.0 dB(A). When placing screens at the source and receiver street canyon edges, much larger effects were found. Although Case C6 is a (geometrical) combination of Cases C4 and C5, the acoustic effect is larger than the simple addition of the separate effects, i.e., it results in an average noise reduction of 3.0 dB(A). Additional calculations with rigid barriers (see Figure 8.1 for Case C6) showed that vegetation is essential for this configuration to obtain a significant noise reduction in the courtyard. Figure 8.6 shows how the Case C6 insertion loss increases with frequency where the data points show averaged values between 30 km/h, 40 km/h, 50 km/h, 60 km/h, and 70 km/h, and the vertical bars represent standard deviations over the various speeds.

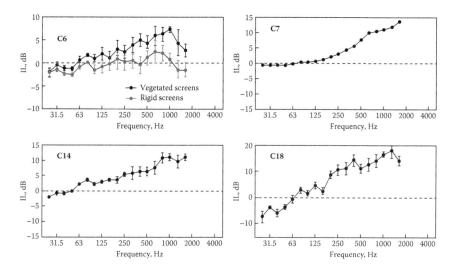

Figure 8.6 Spectrum of insertion losses for selected configurations.

When halving the canyon width, the values presented above are larger by 0.7 dB(A). An additional noise reduction of about 0.5 dB(A) can be obtained for elongated courtyards.

8.4.5 Case C: vegetated courtyard T = Roofs

Changing the roof material to a vegetated roof increases absorption of noise. Roof vegetation applied to flat and pitched roofs and four roof types are considered. The extent to which roof vegetation reduces noise in the courtyard depends on the roof shape. For flat vegetated roofs (C7), the acoustical benefits are 3.0 dB(A). Pitched vegetated roofs (C8, C9) lead to much higher losses, up to 8.0 dB(A) for the double-pitched roof (C10). However, pitched rigid roofs perform worse than flat rigid roofs.

Results of the spectral insertion loss show no effect at very low frequencies, while at higher frequencies very large noise reductions could be obtained. In terms of receiver position, variation in insertion loss is very limited; some variation occurs only at higher frequencies. Lower noise reduction is obtained for elongated courtyards, typically by 0.5–1 dB(A).

8.4.6 Case C: combination of (nonflat) roof shape and vegetated roof

More complex roof shapes also have been considered (see Cases C11–C15). Without vegetation, some of these, C11, C12, and C14, perform worse when compared with roofs of similar shapes, i.e., C8, C9, and C10. The presence of vegetation improves the efficiency of the shaped roofs relative to the flat roof. In particular, the saw-tooth roof (C14) in combination with green roof outperforms flat rigid roofs with more than 6.0 dB(A) and, thus, might be of interest for new buildings. Cases C13 and C15, which are both with depressed geometry, show smaller green roof effects. The reason for this is that the sound wave has less contact with the green roof and the overall effect is mainly due to the roof shape.

Similar to the roofs of typical shapes, noise reduction from the green roofs with novel shapes exhibits low dependence on the receiver position.

8.4.7 Case C: combination of treatments

Combinations of various treatments using vegetation to reduce noise in the courtyard were also studied (see Cases C16–C19). The combined treatments included a combination of roof treatments with low barriers on the edges of the buildings and a combination of vegetated roofs with façade treatments. The combined effect in dB(A) was found to be smaller than a simple sum of individual treatments with green measures. For example, in the case of the combination of green roofs on all buildings and green walls on all the façades (C19), it can be noted that insertion loss in this case is smaller compared to individual cases with green roofs (C7). This is caused by the fact that both green measures are most effective at high frequencies and, by adding more sound-absorbing materials, the other frequencies become more dominant. However, in Case C18, an additional effect of a low-height screen is demonstrated. It increases noise reduction by more than 2.0 dB(A) compared to the case with green roofs, vegetated barriers, and green walls alone, indicating the value of the roof screens in the abatement of noise.

8.4.8 Cases D and E: vegetated opening to courtyards

Vegetating openings to courtyards lead to a reduction of the noise levels in the courtyard of up to 4–5 dB(A) for a 3-m-wide opening both for the opening facing the cross street (D1, D2) and the traffic street (E1, E2). For the cases with a 3-m-high opening, the noise reduction due to vegetation is smaller compared to an opening with a height equal to that of the building (19.2 m), but the difference is less than 1.0 dB(A).

In all cases, the largest decrease in noise levels is found for the highest frequencies and at receiver positions close to the opening. This mitigation measure is efficient, as it needs only a limited amount of vegetated surface area.

8.5 SUMMARY

Vegetation applied to building façades and roofs, and as stand-alone barriers, can, to some extent, reduce noise in urban spaces. However, various aspects related to the geometry of the space, road traffic character, and receiver/listening positions have to be taken into account when considering urban design with vegetation. Acoustic characteristics of plants and growing media (soil or substrate) also are important parameters. Noise mitigation using vegetation, with typical substrate thicknesses of about 0.2 m, is mainly effective at higher frequencies due to the sound absorption processes.

Treatment of all building façades of trafficked streets with green wall may give an average noise reduction of about 1 dB(A) for receivers located along building façades and at 1.5 m height in the streets, while insertion of a vegetated low barrier on central reservations between traffic lanes may give an additional reduction of 2 dB(A). In the case of squares with a trafficked street on one side, a noise reduction of 1 dB(A) can be achieved if green walls are applied to all façades, and screening the square from an adjoining trafficked street with a low-height vegetated barrier may result in a reduction of 4.5 dB(A).

Greening courtyard façades results in average noise reduction in the courtyard of 2 dB(A). This effect is larger at the highest frequencies and for lower receiver positions. Considering the harder reference conditions, namely, with an absorption coefficient of 0.1, an additional gain of 2 dB(A) in noise reduction is predicted.

Placing low-height barriers at both sides of the central building separating a trafficked street from the courtyard results in an average noise reduction of 3 dB(A). Applying green roofs may positively affect the courtyard sound field, with the largest noise reduction of 8.0 dB(A) achieved for a nonflat roof (relative to the same roof shape, but fully rigid). Roof treatments generate reductions that are rather independent of the position of the receiver in the courtyard (along the façades and for receivers along the courtyard floor).

From the investigated combination of treatments, a flat green roof combined with roof barriers has a high potential for providing noise reduction of about 4 dB(A). Combination of green roofs on all buildings with vegetated walls on all building façades in the trafficked street and the courtyard and with low height vegetated barriers on the edges of all buildings results in average noise reduction of 5.0 dB(A), though this would not be the most cost-effective solution to treat noise in the courtyard.

Vegetating openings to courtyards lead to a reduction of noise levels of up to 4–5 dB(A) for a 3-m-wide opening both for the opening facing the cross street and the trafficked street.

Predictions of the ability of vegetation to reduce noise in urban spaces largely depend on the absorption properties of the underlying materials used in the calculations for reference configurations without vegetation. For example, decreasing the frequency-independent absorption coefficient from 0.33 to 0.1 for the reference cases results in increase of noise reduction of 2 dB(A) in the street with vegetated façades and of 2–3 dB(A) in the square with vegetated façades.

REFERENCES

1. Barrio, E. P. D. 1998. Analysis of the green roofs cooling potential in buildings. *Energy and Buildings* 27(2): 179–193.
2. Takebayashi, H., and M. Moriyama. 2007. Surface heat budget on green roof and high reflection roof for mitigation of urban heat island. *Building and Environment* 42(8): 2971–2979.
3. Stovin, V., G. Vesuviano, and H. Kasmin. 2012. The hydrological performance of a green roof test bed under U.K. climatic conditions. *Journal of Hydrology* 414: 148–161.
4. Yang, H. S., J. Kang, and M. S. Choi. 2012. Acoustic effects of green roof systems on a low-profiled structure at street level. *Building and Environment* 50: 44–55.
5. Fernandez-Canero, R., and P. Gonzalez-Redondo. 2010. Green roofs as a habitat for birds: A review. *Journal of Animal and Veterinary Advances* 9(15): 2041–2052.
6. MacIvor, J. S., and J. Lundholm. 2011. Insect species composition and diversity on intensive green roofs and adjacent level-ground habitats. *Urban Ecosystems* 14(2): 225–241.
7. Van Renterghem, T., and D. Botteldooren. 2008. Numerical evaluation of sound propagating over green roofs. *Journal of Sound and Vibration* 317(3): 781–799.
8. Van Renterghem, T., M. Hornikx, J. Forssen, and D. Botteldooren. 2013. The potential of building envelope greening to achieve quietness. *Building and Environment* 61, 34–44.
9. ISO. 1998. *Acoustics—Determination of sound absorption coefficient and impedance in impedance tubes. Part 2: Transfer-function method (ISO 10534-2).* Geneva: International Standards Organization.
10. ISO. 2003. *Acoustics—Measurement of sound absorption in a reverberation room (ISO 354:2003).* Geneva: International Standards Organization.
11. Van Renterghem, T., D. Botteldooren, J. Kang, and H.-S. Yang. 2014. Acoustical characteristics of trees, shrubs and hedges. In *Environmental methods for transport noise reduction*, eds. M. E. Nilsson, R. Klæboe, and J. Bengtsson. (pp. 81–92) Oxford, U.K.: Spon Press.
12. Yang, H.-S., J. Kang, and C. Cheal. 2013. Random-incidence absorption and scattering coefficients of vegetation. *Acta Acustica United with Acustica* 99(3): 379–388.
13. Kang, J. 2007. *Urban sound environment.* London: Taylor & Francis.
14. Kang, J. 2002. Numerical modelling of the sound fields in urban streets with diffusely reflecting boundaries. *Journal of Sound and Vibration* 258(5): 793–813.

15. ISO. 2007. *Acoustics—Attenuation of sound during propagation outdoors—Part 2: A general method of calculation (ISO 9613-2).* Geneva: International Standards Organization.
16. Jonasson, H. 2007. Acoustical source modelling of road vehicles. *Acta Acustica United with Acustica* 93(2): 173–184.
17. Botteldooren, D. 1995. Finite-difference time-domain simulation of low-frequency room acoustic problems. *Journal of the Acoustical Society of America* 98: 3302–3308.
18. Hornikx, M., and J. Forssén. 2007. The 2.5-dimensional equivalent sources method for directly exposed and shielded urban canyons. *The Journal of the Acoustical Society of America* 122: 2532.
19. Van Renterghem, T., and D. Botteldooren. 2003. Numerical simulation of the effect of trees on downwind noise barrier performance. *Acta Acustica United with Acustica* 89: 764–778.
20. Salomons, E. M., R. Blumrich, and D. Heimann. 2002. Eulerian time-domain model for sound propagation over a finite-impedance ground surface. Comparison with frequency-domain models. *Acta Acustica United with Acustica* 88(4): 483–492.
21. Hornikx, M., R. Waxler, and J. Forssén. 2010. The extended Fourier pseudospectral time-domain method for atmospheric sound propagation. *The Journal of the Acoustical Society of America* 128: 1632.
22. Meng, Y., and J. Kang. 2007. Combined ray tracing and radiosity simulation for urban open spaces. Paper presented at the *Proceedings of 19th International Congress on Acoustics*, International Commission for Acoustics (ICA), September 2–7, Madrid.
23. Horoshenkov, V., C. Hothersall, and E. Mercy. 1999. Scale modelling of sound propagation in a city street canyon. *Journal of Sound and Vibration* 223(5): 795–819.
24. Ismail, M., and D. Oldham. 2005. A scale model investigation of sound reflection from building façades. *Applied Acoustics* 66(2): 123–147.
25. Hornikx, M., and J. Forssén. 2008. A scale model study of parallel urban canyons. *Acta Acustica United with Acustica* 94(2): 265–281.
26. Öhrström, E., A. Skånberg, H. Svensson, and A. Gidlöf Gunarsson. 2006. Effects of road traffic noise and the benefit of access to quietness. *Journal of Sound and Vibration* 295: 40–59.
27. de Kluizenaar, Y., E. M. Salomons, S. A. Janssen, F. J. van Lenthe, H. Vos, H. Zhou, H. M. Miedema, and J. P. Mackenbach. 2011. Urban road traffic noise and annoyance: The effect of a quiet façade. *The Journal of the Acoustical Society of America* 130: 1936.
28. Van Renterghem, T., and D. Botteldooren. 2012. Focused study on the quiet side effect in dwellings highly exposed to road traffic noise. *International Journal of Environmental Research and Public Health* 9(12): 4292–4310.

Chapter 9

Perceptual effects of noise mitigation

Mats E. Nilsson, Dick Botteldooren, Jin Yong Jeon,
Maria Rådsten-Ekman, Bert De Coensel, Joo Young
Hong, Julien Maillard, and Bruno Vincent

Abstract. Noise mitigation reduces the audibility of a noise source at the location of the receiver, making the source less annoying and less likely to interfere with activities, such as sleep, rest, and speech. Many mitigation methods change temporal and spectral properties of noise, which may influence perceived annoyance, over and above the effect related to the overall reduction in A-weighted sound pressure level. Noise reduction also may increase the noticeability of other sources, which may influence the perception of the overall acoustic environment. Finally, well-designed noise mitigation solutions may improve the visual environment, e.g., a vegetated noise barrier or earth berm can visually shield the traffic and increase the amount of visible greenery. This chapter provides examples of such perceptual effects of noise mitigation, from effects on perception of the noise itself, via effects on the soundscape, to potential effects on the overall audio-visual environment.

9.1 INTRODUCTION

The goal of noise mitigation is to reduce the audibility of a noise source at the location of the receiver, making the source less annoying and less likely to interfere with activities, such as sleep, rest, and speech. Noise mitigation also can indirectly influence the acoustic environment (soundscape) by making previously masked sounds, such as birdsong or sounds of moving water, more noticeable. In addition to auditory effects, noise mitigation may improve the scenery of a place, e.g., a vegetated noise barrier or earth berm can visually shield the traffic and increase the amount of visible greenery. This chapter provides examples of such perceptual effects of noise mitigation, from effects on perception of the noise itself, via effects on the soundscape, to potential effects on the overall audio-visual environment.

The first section of the chapter discusses the psychoacoustics of noise mitigation, with a primary focus on how well accomplished reductions in sound pressure level predict changes in perceived annoyance of noise. The second section broadens the perspective and looks not only at unwanted sounds,

195

but at the overall soundscape, including its wanted and unwanted components. The main focus in this section is on the role of sound masking and how it relates to noise mitigation. The last section broadens the picture further and looks at the total environment, and, in particular, audio-visual aspects of noise mitigation.

This chapter is limited to *perceptual* aspects of noise mitigation, and will not discuss long-term effects of noise, such as effects on residential noise annoyance, sleep disturbance, or cardiovascular effects related to changes in exposure caused by noise mitigation.[1-3] Perceptions are immediate, but their effects may be long term. Noise mitigation that makes the noises less loud and less disturbing will, of course influence how the noise source is assessed in the long run, for instance, in questionnaire studies, where residents typically are asked to evaluate noise sources as experienced at home during the past several months.[4]

9.2 NOISE: PSYCHOACOUSTICS OF NOISE MITIGATION

Noise mitigation reduces the audibility of a noise source by reducing the sound pressure level of the noise at the point of the receiver. The efficiency of noise-mitigation methods is typically assessed in terms of the achieved reduction in A-weighted sound pressure level, expressed in dB(A), which, in many cases, gives a fair indication of the effect on the audibility of noise.

However, most mitigation methods do not merely reduce the overall level, but also alter other aspects of the noise. For example, a noise barrier will reduce high frequency sounds to a larger extent than low frequency sounds, and will reduce the temporal variability of the noise behind the barrier. Such changes may influence how annoying the noise is perceived to be, over and above the effects explained by the overall reduction in A-weighted sound pressure levels.

Spectral or time-pattern changes may influence perceived annoyance in two ways. First, the perceived *loudness* of the noise may change in a way not well predicted by the reduction in A-weighted sound pressure levels. For most noise sources, including traffic, perceived loudness is the main determinant of annoyance.[5] That is, the louder the noise, the more annoying it is perceived to be. A-weighting has been shown to underestimate the loudness of traffic noise with sizeable low-frequency components.[6,7] For noise mitigation methods that reduce high frequency sounds more than low frequency sounds, the reduction in A-weighted sound pressure levels, therefore, may overestimate the reduction in perceived loudness, and, thereby, annoyance. An alternative to the A-weighted sound pressure level is the psychoacoustic indicator loudness (unit sone) and loudness level (unit phon).[8,9] This indicator was derived to predict the perceived loudness of sounds, and it is based on a detailed model of how sound is processed in the ear (i.e., the peripheral hearing system). Loudness level[9] was found to be a better indicator of annoyance than the A-weighted sound pressure level in an experimental study of the perceptual effect of a 4.1-m-high barrier close to a highway.[10]

Secondly, noise mitigation may change the perceived character of the noise and, thereby, have an effect on annoyance in addition to effects related to the loudness of the noise. Two equally loud sounds may differ in perceived annoyance due to differences in their perceived character or quality. The importance of the character of the noise for its annoyance seems to be inversely related to the loudness of the sound; two moderately loud sounds are more likely to differ in annoyance than two very loud sounds.[5]

Certain noise characteristics are especially annoying, for example, tonal components (e.g., ventilation system noise), intermittent time patterns (e.g., pile driving), slow and regular fluctuating time patterns (e.g., wind turbine noise), rapid fluctuations leading to a rough sound (e.g., drilling noise), and strong, high-frequency components leading to a sharp sound (e.g., squeals from rails). Strong, low-frequency components may be perceived as specifically annoying and tiring, especially when they interact with resonances in the building shell (e.g., noise from an idling diesel engine). Moderate hearing loss (e.g., age related) may strengthen the importance of some of these characteristics.

9.2.1 Case study of low, vegetated barriers

The perceptual effects of a low, vegetated noise barrier in central Lyon, France, were evaluated in a field questionnaire study, complemented with a listening experiment in the laboratory. The barrier was erected to protect a popular esplanade from road traffic noise. The purpose of the evaluation was to determine the acoustic and perceptual effects of the barrier.

In the field study, pedestrians were asked to assess the sound environment. Questionnaire responses were collected on two occasions: one before and one after the barrier was erected. Each time, data were collected at two locations: at a place behind the barrier and a place 20 m to the side of the barrier (Figure 9.1(a)(b)). Acoustic measurements made at the same locations where the questionnaire was completed showed that the sound pressure level behind the barrier was, on average, 4 dB(A) lower than without the barrier. The noise variability also was reduced by the barrier, whereas the relative level of low-frequency sound increased, because barriers reduce high-frequency sounds more than low-frequency sounds (Figure 9.2(a)–(c)). These measurements illustrate that noise barriers not only influence the overall level, but also spectral and time-pattern parameters.

The questionnaire responses showed that the barrier improved the perceived acoustic environment. The percentage of noise-annoying respondents decreased from 59% at places uninfluenced by the barrier to 47% behind

(a)

(b)

Figure 9.1 Questionnaire data collection and simultaneous sound level measurements behind (a) or beside (b) a vegetated barrier in central Lyon, France.

the barrier, and ratings of the overall quality of the sound environment indicated that the barrier made the soundscape slightly calmer and less unpleasant. However, traffic was still the dominant sound source, which explains the fairly high number of annoyed respondents also after mitigation.[11]

A listening experiment with traffic noise events simultaneously recorded behind and beside the barrier verified that the barrier reduced the

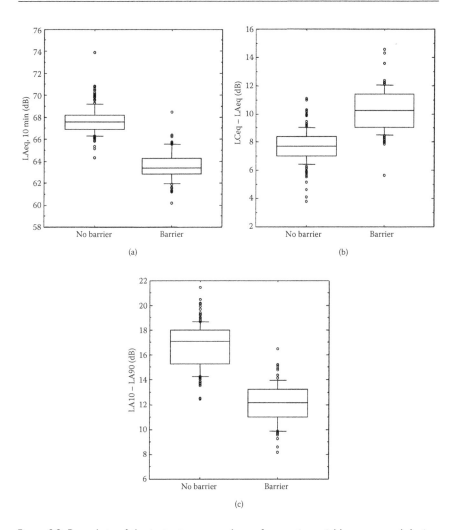

Figure 9.2 Box plots of the instantaneous values of acoustic variables measured during time intervals when questionnaire data were collected. "No barrier" measurements were either taken before the barrier was erected or after, at the side of the barrier; "barrier" measurements were taken behind the barrier. (a) A-weighted sound pressure level, LAeq, 10 min. (b) difference between C- and A-weighted sound pressure level (LCeq, 10 min–LAeq, 10 min). (c) Difference between A-weighted sound pressure levels exceeded 10% and 90% of the time (LA10–LA90). The five vertical lines of each box plot show, from top to bottom, the 10th, 25th, 50th (median), 75th, and 90th percentile of the data; the circles show data points more extreme than the 10th or 90th percentile.

annoyance of the traffic noise, and that this effect was fairly well predicted by the associated reduction of A-weighted sound pressure level (Figure 9.3(c)). However, there was a tendency for the annoyance reduction to be a little less than would be expected from the A-weighted sound pressure level reduction (green symbols located slightly above grey symbols at similar A-weighted sound pressure levels). The loudness level was a slightly better predictor of noise annoyance (Figure 9.3(d)) than the A-weighted sound pressure level, but had the same trend, which suggests that part of the effect was related to perceived loudness and part to the perceived character of the noise. Statistical analysis suggested that this could be explained partly by the barrier's lower reduction of low-frequency components of the sound, as compared to high-frequency components, witnessed by a larger difference between C- and A-weighted sound pressure levels compared to without barrier influence. The barrier also reduced the noise variability, measured as the difference between levels exceeding 10% and 90% of the time. However, statistical analyses suggested that this did not strongly influence the perceived annoyance of the noise.

9.2.2 Perceptual effects of soft and hard ground along tramways

The acoustic and perceptual effects of soft or hard ground between tramways and receivers were evaluated in a study involving measurements, binaural recordings, and a listening experiment. Recordings at a height of 1.5 m were made 4 and 7 m from a tramway in Grenoble, France, at a location with soft ground (grass) and at another location with hard ground (asphalt). A large number of tramway passages were recorded, and these were matched to allow comparisons of recordings made in places bordered by different types of ground, but of trams of the same type travelling at the same speed (Figure 9.4(a)(b)).

At the closer distance (4 m), sound pressure levels from tram passages were about the same at both the grass and asphalt locations. However, at a distance of 7 m, the grass reduced the level of noise by approximately 3 dB(A) compared with the asphalt (cf. this volume Chapter 6[12]).

A listening experiment using segments of tram noise centred around the maximum level verified that recordings made near the tramway were about equally annoying, regardless of whether the tramway was bordered by grass or asphalt; for recordings made farther from the tramway, however, the grass margin clearly resulted in less annoyance. The effect at this distance could be predicted fairly well from the associated A-weighted sound pressure level reduction. There was, however, a clear tendency for the annoyance difference between the grass and asphalt recordings to be greater than one would predict from the A-weighted sound pressure level difference alone (Figure 9.5(a)). Verbal reports from listeners suggest that a main perceptual effect was a reduction of high-frequency sounds in the tram noise,

(a)

(b)

Figure 9.3 Recordings made using artificial head technology (a) to obtain high-quality recordings for listening experiments (b).

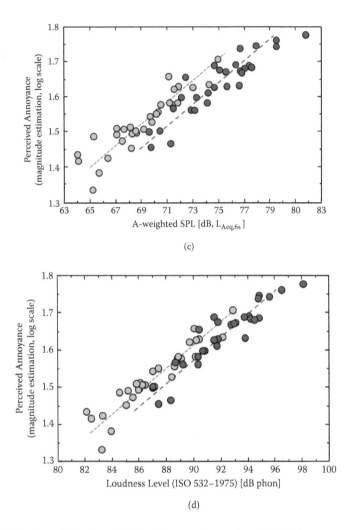

Figure 9.3 (Continued) Results of a listening experiment in which 30 listeners assessed the perceived annoyance of traffic noise events recorded at barrier-shielded (green circles) or non-barrier shielded (grey circles) places. The y-axis shows perceived annoyance (magnitude estimates on a log scale) and the x-axis shows A-weighted sound pressure level (c) or loudness level (d).

and this observation was supported by acoustic analyses (see spectra in Figure 9.4(c)).

This is an example of noise mitigation that reduced not only the perceived loudness, but also altered the character of the noise in a way that led to a larger reduction in perceived annoyance than expected

(a)

(b)

Figure 9.4 Photos from recordings in Grenoble of tram noise, at a location with asphalt (a) or grass (b) close to the rail and microphone.

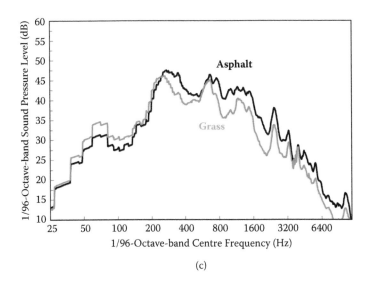

(c)

Figure 9.4 (Continued) (c) Average 1/96-octave band frequency spectra for experimental sounds recorded at 7 m distance from the tram at a location with asphalt (black line) or grass (green line) between source and receiver.

from the reduction in A-weighted sound pressure level or loudness level. The effect was fairly large (1–2 dB(A)) and statistically significant, and corresponds to the horizontal distance between the regression lines in Figure 9.5(a) and (b).

The results of this experiment suggested that replacing hard ground with soft ground (grass) between tramways and listeners may reduce sound pressure levels by about 3 dB(A) at a 7-m distance, and that the associated effect on perceived annoyance may be even greater thanks to perceptual changes related to the change in spectral composition of the noise. Further studies are needed to confirm these results, and to find out whether they generalise to other settings and noise sources. Nevertheless, the results are promising and this is, to the best of our knowledge, the first experimental study that has demonstrated a case where noise mitigation might lead to a perceptually larger effect than expected from the reduction in overall level (the opposite to the previous example; see Figure 9.3(c)(d)). Note that this is not the same effect as has been demonstrated in several field studies, where noise reduction has been followed by an initial, temporary strong reduction in long-term noise annoyance at home.[13] Such effects are thought to be related to attitudinal or response behaviour factors, rather than to purely perceptual factors as in this experiment (where the listeners were blind to what condition they were assessing).

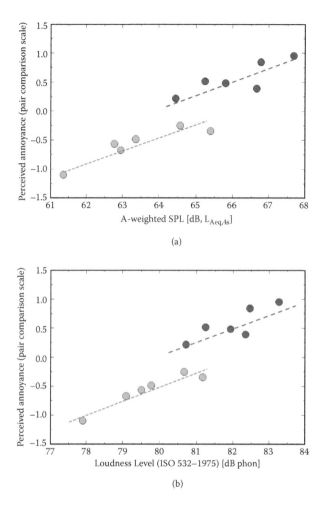

Figure 9.5 Results of a listening experiment in which 29 listeners assessed the perceived annoyance of tram passages recorded at a point 7 m from the track, bordered by soft (green) or hard (black) ground. The y-axis shows perceived annoyance (scale derived from pair-wise comparisons using Thurstonian scaling[13]) and the x-axis shows A-weighted sound pressure level (a) or loudness level (b) of the tram noises.

9.3 SOUNDSCAPE: WANTED AND UNWANTED SOUNDS IN INTERACTIONS

Conventional noise control aims at protecting people from harmful effects of noise. During the past few decades, the primary focus of noise control has been the indoor environment of residential dwellings, with the goal to protect the indoor environment from external (e.g., road traffic) or internal noises (e.g., ventilation), and thereby minimise annoyance, sleep disturbance, and other adverse effects of noise. When successful, a quiet indoor environment is created, and residents are free to design their acoustic environment as they like, by turning on the radio, playing music, or just keeping it as it is.

Outdoors, absolute quietness is neither possible nor desirable. We expect the outdoor soundscape to contain sounds from many sources. Specific sources can often be heard individually; for example, in a city park, you may hear sounds from traffic, birds, and wind in the trees at the same time. This complexity notwithstanding, we can easily assess the overall quality of the acoustic environment (the soundscape), and say whether we prefer one soundscape over another, or say whether we like or dislike a given soundscape as perceived in relation to our goals and activities (rest, relaxation, enjoyment, etc.).

The overall quality of the soundscape (on a like–dislike dimension) depends on its composition of specific sounds and on how these sounds are perceived in interaction. The sounds of a specific soundscape may roughly be classified as unwanted, wanted, or neutral sounds. Unwanted sounds (noise) detract and wanted sounds add to the overall quality of the soundscape, whereas neutral sounds neither do good nor harm.

Obviously, the same sound may be preferred in one environment but not in another. For example, traffic noise may add to the vibrancy of a downtown shopping area, but would be a nuisance in a green city park intended for rest. Moreover, people may differ in which sounds they prefer and dislike in a given environment, although, in general, people tend to agree more than they disagree about which sounds they prefer.[14,15] For example, in urban open spaces, most people classify birdsong and water-generated sounds as wanted, and technological sounds, such as traffic noise and ventilation noise, as unwanted.[16,17]

9.3.1 Auditory masking and noticeability

Successful noise mitigation will improve the soundscape by increasing its ratio of wanted to unwanted sound. This is achieved not only by reducing unwanted sounds (noise), but, potentially, also by allowing wanted sounds to attract attention once the "masking" noise has been removed or attenuated. A deeper understanding of the effect of noise mitigation on soundscape quality, therefore, requires knowledge on how wanted and unwanted sounds are perceived in interaction.

Researchers at Ghent University have developed a theoretical and computational framework for studying such interactions. At the heart of their approach is their notice-event model,[18] which takes its starting point in the following assumptions:

1. Only sounds that are both audible and noticed (through a notice event) may be annoying and potentially threatening to well-being.

2. A notice event is any event that draws attention to the sound. This could occur when a sound suddenly emerges above the background generated by everyday activities, e.g., when the background level itself drops, when attention for one's environment increases, etc.

Whether or not a sound is audible and noticed in a given environment is determined by peripheral and central processes of the auditory system. At the periphery, a masking sound makes a target sound inaudible (complete masking) or less loud (partial masking) by decreasing signal-to-noise ratios (SNRs) in the frequency regions surrounding the target at the basilar membrane of the inner ear.[19] This kind of masking is referred to as energetic masking, to distinguish it from effects on a sound's noticeability related to auditory processes at higher levels, sometimes called *informational* masking.[20,21]

Energetic masking is asymmetric in the sense that low-frequency sounds mask high-frequency sounds more than the other way around. Many unwanted sounds, like traffic noise, have more energy in the low-frequency part of the spectrum than typical wanted sounds, such as sounds from water features.[22–24] Consequently, noise reduction may increase audibility of wanted sounds previously masked, provided that also low frequencies are reduced.

Whether an audible sound is noticed is determined largely by attention mechanisms. An inattentive visitor to a space will notice some of the sounds without noticing others. Acoustic design aiming to provide a restorative environment for people using the space may include facilitating the occurrence of notice events for wanted sounds. Making wanted sounds more noticeable requires a detailed analysis of the spectro-temporal structure of both the wanted and unwanted sounds.

Note that the noticeability of sounds is not only influenced by the amount of energetic masking and auditory attention. Prior knowledge and expectations of what kind of sounds to hear in a given environment, as well as information from other senses, may enhance the noticeability of sound sources. For example, if the peak noise level is low enough, noticing might be influenced by the sound source being visible or not. Visible sound sources, particularly when moving, can attract attention visually and, thereby, direct auditory attention to a source that would otherwise have gone unnoticed.

In listening studies, participants are instructed to assess sounds in various ways. The instructions, thereby, direct the listener's auditory attention to specific aspects of the soundscape, and, therefore, it is difficult to study noticeability using such methods. Accordingly, listening studies are used mainly to assess perceived loudness, annoyance, or other characteristics of specific sounds or soundscapes (examples were given in the previous section). An alternative approach is to model the relevant phenomena and to use these models to compute a sound's noticeability. The notice event model discussed above (Figure 9.6) has been implemented in computer models predicting whether the average visitor will notice a sound in a given soundscape, and developments of the model are increasingly accurate and powerful.[18,25]

Figure 9.7 gives an example of noticeability computation predicting time intervals in which attention is attracted by sounds from birdsong in the presence of traffic noise, in a given environment with and without a noise barrier. In this 30-second time interval, birdsong is predicted to be noticed three times more often when a noise barrier (reducing the noise level by 6 dB(A)) is introduced.

This illustrates the potentially dual effect of noise mitigation, reducing unwanted sounds and increasing the noticeability of wanted sounds. This naturally leads to the idea discussed next, of adding wanted sounds to improve the soundscape by masking unwanted sounds.

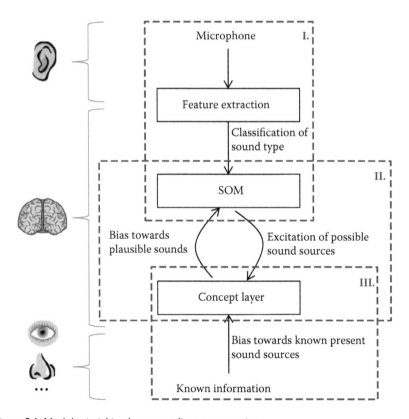

Figure 9.6 Model mimicking human auditory perception.

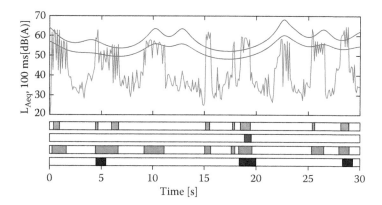

Figure 9.7 (Top) Fragment of birdsound (green) combined with traffic sound recorded at 15 m distance (red) and traffic noise behind a barrier (blue); (bottom) the pink and light blue strips indicate time intervals in which birdsound is not energetically masked; the red and dark blue strips indicate noticed birdsound.

9.3.2 Adding wanted sounds

Is it possible to improve the soundscape by adding wanted sounds? In theory, yes. Adding wanted sounds may mask unwanted sounds or divert attention away from unwanted to wanted sounds. In practice, however, this is easier said than done, for reasons discussed below.

Energetic masking of unwanted by wanted sounds is difficult to achieve for at least three reasons. Firstly, the frequency asymmetry of masking will play in favour of the unwanted sounds, which typically contain more low-frequency energy than many wanted sounds. This was illustrated in a study that measured perceived loudness of traffic and fountain sounds recorded in a city park, and found that the traffic noise, which had sizeable low-frequency components, reduced the loudness of the fountain sound, dominated by high-frequency components, considerably more than the other way around.[23] Galbrun and Ali[24] looked at different types of water features, and found that only waterfall sounds contained sizeable low-frequency components that might effectively mask road traffic noise. Unfortunately, waterfall sounds are among the least pleasant water-generated sounds, and, therefore, might not help to improve the overall soundscape. In fact, recent experimental studies have shown detrimental effects of adding waterfall sounds on assessments of overall soundscape quality.[26,27]

Secondly, differences in temporal variability between masker and target sound will diminish any masking effects. Even if a noise is masked at periods when its level is low in comparison to the masking sound, the noise may be audible and thereby potentially noticeable and annoying at periods when its level is high. This was demonstrated by De Coensel et al.[28] who found sounds from fountains reduce perceived loudness of freeway noise (low variability) more than noise from a major road (moderate variability), but had no effect on the loudness of noise from a minor road (high variability).

Thirdly, when sound sources are located in different directions from the listener, binaural localization cues will make it easier to hear the sounds separately than when they are located in the same direction. This is known as spatial unmasking and is related to how sounds at different locations contribute to binaural level and time differences.[29] If a fountain is located to the left of a listener, the level of the fountain sound at the listener's right ear will be reduced by the acoustic shadow of the head. Noise from a road located in front of, or to the right of, the listener will be less shadowed at the right ear, and, by attending to this ear, the listener may hear the road traffic noise better than if it had come from the same direction as the fountain sound. In addition, the auditory system compares the phase of signals between ears and this information is used to "binaurally unmask" sounds from different locations. It, therefore, is advisable to place a masking sound source (e.g., a fountain) in the same direction from the intended listener as the target source (e.g., road), for example, by placing the fountain in-between the listener and the road.

Even if wanted sounds do not easily mask unwanted sounds energetically, there may still be an effect on noticeability. For example, birdsong will hardly energetically mask road traffic noise,[28,30] because of birdsong's high frequency content, intermittent temporal pattern, and the elevated location of the source. However, birdsong may still attract attention away from the traffic noise and thereby reduce it noticeability. The notice event model can be used to explore this possibility, as illustrated below.

Oldoni et al.[25] used an updated version of the notice event model to explore the perceptual effects of attracting songbirds at an urban location. Once the model is trained for a particular location, it classifies the sounds that are present in the soundscape and simulates how a typical listener would switch attention over time between different sounds. The model thus allows to assess the perceptual effect of introducing additional sounds.[25]

For the particular application described here, a fixed sound measurement station was installed in the city of Ghent, Belgium. The soundscape at the chosen location consisted mainly of a mixture of road traffic noise due to private and public transport on a nearby urban road, and the noise from pedestrians due to the proximity of several shops and one educational institution. In a first stage, a 1/3-octave band spectrum at 1-s time intervals was measured during three weeks at this location. These measurement data were then used to train the computational model, such that it would be able to classify the typical sounds that can be heard at the particular location over the course of the day.

In a second stage, a 1-h sound recording was performed at the location; the L_{Aeq} during this period was 68 dB(A). Subsequently, a series of 30 artificial 1-h soundscapes were created by mixing the original recording with an increasing number of birdsounds at random instances in time. For this, a series of bird vocalizations without background noise, with a duration of up to a few seconds, was used, for which the peak level was adjusted to match the peak level of the few birdsounds present in the original recording. The 1-h L_{Aeq} of the added birdsound ranged from 46 dB(A), representing a few sporadic vocalisations, to 76 dB(A), representing a quasi-continuous bird chorus, resulting in an SNRs for birdsound versus background ranging from −22 dB to +8 dB.

The computational model, trained for the particular microphone location, was used to classify the sounds that are present in the 30 artificial soundscapes. In the audibility analyses (not taking auditory attention into account), it was found that the model clearly distinguished between the sound of individual birds and the sound of a bird chorus. Figure 9.8(a) shows the percentage of the time that individual bird chirp is dominant, which increases monotonically with SNR until a peak is reached at an SNR equal to −2 dB. At that point, the percentage of the time that bird chorus dominates the soundscape starts to increase with increasing SNR.

The noticeability analyses used the same procedure, but now also taking into account auditory attention mechanisms (see Figure 9.8(b)). The output of the model can then be interpreted as the fraction of time each particular sound is noticed (receives attention). It can be seen that for lower SNR, the percentage of time that attention is paid to birds is slightly higher than in Figure 9.8(a), while for higher SNR, this percentage is lower. This is indeed the expected behaviour, as for lower SNR, each time birdsound is detectable it will get attention because its saliency is higher than the background, and the particular sound source may even receive attention even for a short time after the sound has actually stopped, because of expectation and focussed listening. For higher SNR, birdsound will be continuously detectable, and the inhibition-of-return mechanism implemented in the model will cause attention to shift away. Considering that sounds need to be noticed, in order to contribute to the appraisal of a soundscape, these results are in accordance with empirical results showing the potential of birdsound to increase soundscape pleasantness and eventfulness.[28]

In summary, masking unwanted sounds (noise) with wanted sounds is in most situations difficult to achieve. Studies using water-generated sounds as maskers show small, if any, effects on traffic noise loudness, and, in some conditions, detrimental effects on overall soundscape quality.[27] However, noticeability of noise may be reduced, as suggested by results from models of auditory attention of road traffic noise combined with twittering of birds.

All in all, it seems fair to conclude that unwanted sounds (noise) do more harm to the soundscape than wanted sounds do good. Therefore, noise mitigation will always be the main method for soundscape improvement.

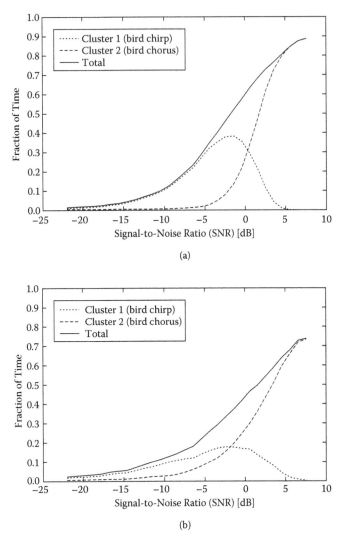

Figure 9.8 Fraction of time birdsong is audible (a) and receives attention (b), as a function of signal-to-noise ratio (SNR) between birdsound and background noise. The dotted line denotes individual birds.

However, soundscape design based on adding sounds may be used as a complement to noise mitigation, especially in environments with moderate-to-low levels of noise. Greening will add natural sound, both vegetation-generated sounds and, to some degree, wildlife sound, and some additional improvement in soundscape quality could be expected from using natural materials for noise mitigation.

9.4 ENVIRONMENT: AUDIO-VISUAL INTERACTIONS

Many noise-mitigation methods influence the visual environment as well, and the use of vegetated mitigation elements can improve the visual quality of environments. The extent to which such visual changes also influence *auditory* perceptions is unclear, and results in different directions have been published.[31] Any such effects of visual impressions on *auditory* perception are probably small (closing one's eyes doesn't substantially change the auditory perception of a place). More important than auditory perceptions, however, are effects on how the overall environment is perceived, and noise-mitigation methods that, in addition to reducing noise, also improve aesthetic values are obviously better than methods that do not.

Aesthetic values are of particular importance in outdoor areas intended for rest and relaxation, such as city parks or recreation grounds. Previous research suggests that the sound environment and scenery independently contribute to the overall perceived tranquillity of such areas, and that low sound levels combined with a view dominated by vegetation would be associated with a high degree of tranquillity.[32,33] This suggests that noise mitigation that in addition to noise reduction also increases the amount of greenery would be suitable for areas intended for rest and relaxation, by simultaneously increasing visual and auditory tranquillity.

Effects also may be seen on urban streets, as Hong and Jeon[27] showed in an experiment where participants assessed auditory, visual, and audio-visual stimuli on an 11-point preference scale (from 0 = not at all, to 11 = extremely preferred). The visual stimuli were created by manipulating a street view to obtain a set of "streetscapes" with combinations of various vegetation and water features (Figure 9.9). The street was selected based on a previous sound walk study in downtown Seoul,[34] where participants had assessed the location negatively with regard to both its auditory and visual qualities.

The auditory stimuli consisted of one unwanted sound, traffic noise, and three wanted sounds: sound from falling water, stream water, and twittering of birds, presented alone or in combination. The traffic noise was set to either 55 or 70 dB(A) L_{Aeq} and the water and bird sounds to either 58 or 73 dB(A) L_{Aeq}. Preference ratings of the sounds heard alone showed that the traffic noise was clearly the least-preferred sound. Among the three sounds selected to represent wanted sounds, the sound of twittering of birds was most preferred, followed by the sound of streaming water, with the sound of falling water as least preferred (but still much more highly rated than the road traffic noise) (Figure 9.10).

The street seen without vegetation or water features was less preferred than with such features (white bars in Figure 9.11). In the audio-visual conditions, preference ratings were similar to the visual-only ratings when the

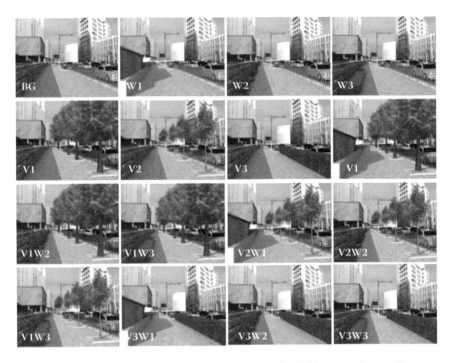

Figure 9.9 Images of streetscapes in downtown Seoul, South Korea, with a combination of vegetation (V) and water features (W).

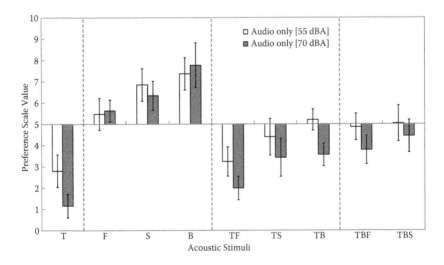

Figure 9.10 Mean preference scores for the acoustic stimuli (audio-only condition). T = traffic noise; F = sound from falling water; S = sound from streaming water; B = birdsong.

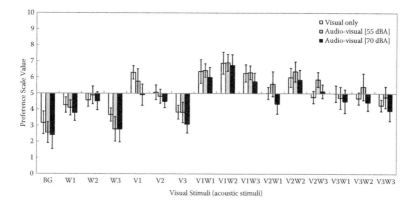

Figure 9.11 Mean preference scores for visual-only (white bars) and audio-visual presentation of streetscapes (grey bars). Error bars indicate 95% confidence intervals.

traffic noise was of a moderate level (55 dB), but, in general, less for high levels of traffic noise (70 dB) (see Figure 9.11).

Hong and Jeon conclude that increases in greenery from trees or bushes can improve streetscapes, but water features as visual components may not significantly improve the perceived view. Among natural sounds for sound-scape elements, bird sounds were more useful for enhancing soundscape quality with road traffic noise than were water sounds. However, it was revealed that water sounds, such as falling water, could decrease the overall quality of the environment when the road traffic noise level was higher than 70 dB(A). Thus, care must be taken to consider these characteristics when designing soundscapes using water sounds.

In a similar experiment, Hong and Jeon[35] evaluated various noise barrier designs, including barriers made of timber, metal, vegetated substrate, concrete, and translucent acrylic. Participants in the experiment were asked to assess audio-visual presentations of barriers with normalised traffic noise levels (55 or 65 dB(A)) (Figure 9.12). Vegetation was found to enhance overall preference, as seen in Figure 9.13. For both the translucent acrylic, Tr, and the concrete barrier, Co, preferences increased from no vegetation to sparse vegetation, (L), to dense vegetation, (H). The highest preference ratings were found for the concrete covered to a high degree with climbing ivy. In this study, it also was found that participants assessed the noise reduction potential of barriers higher when the barriers were covered with vegetation. These results are based on laboratory experiments and need to be evaluated in real-life settings. Nevertheless, the results point to the possibility of using vegetation on barriers to enhance not only the overall preference of a place, but also perceived noise barrier performance.

Figure 9.12 Photomontages for the different types of noise barriers: (a) timber, (b) metal, (c) concrete, (d) translucent acrylic, (e) vegetated substrate, (f) concrete with sparse vegetation cover, (g) concrete with dense vegetation cover, (h) translucent acrylic with sparse vegetation cover, and (i) translucent acrylic with dense vegetation cover.

9.5 CONCLUDING REMARKS

This chapter has provided examples of perceptual effects of noise mitigation, from effects on perception of the noise itself, via effects on the soundscape, to potential effects on the overall audio-visual environment. The examples suggest that conventional noise mitigation, which solely focusses on A-weighted sound pressure level reductions, would benefit from perceptual analysis. For example, perceptual effects of methods that mainly affect the high-frequency part of the spectrum may not be well predicted by the A-weighted sound pressure level, and alternative indicators, for instance, the loudness level, may be used as a complement. In addition to loudness level, spectral shape was shown to significantly affect instantaneous annoyance caused by traffic noise. The potential effect of noise mitigation on the soundscape should always be considered, and methods

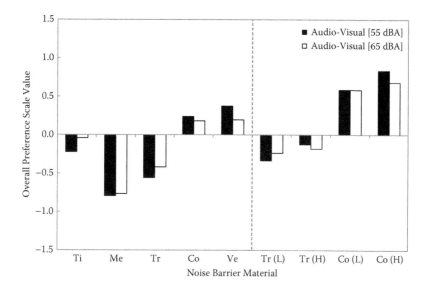

Figure 9.13 Scale value of overall preference for environment according to combined audio-visual stimuli: timber (Ti), metal (Me), translucent acrylic (Tr), concrete (Co), and vegetation (Ve). L and H (in parentheses) denote low and high densities of vegetation cover, respectively. Scale values were derived from results of a pair comparison task (using the Thurstonian scaling model).

that not only reduce the noise but also increase the prevalence of wanted sounds, for instance, vegetated solutions attracting birds and generating sounds from wind in the leaves, may have positive effects on the soundscape, over and above the effects associated with the reduction of noise. Models for predicting the audibility and noticeability of the sounds in the environment are under development, and may prove useful in the future for evaluating effects of noise mitigation on soundscapes. Finally, well-designed, noise-mitigation methods improve visual aesthetic values, and vegetation is a particularly valuable design element for this purpose. Green noise mitigation that reduces noise, increases prevalence and noticeability of wanted sounds, as well as improves the visual environment, could be well motivated also in situations where a simplistic analysis solely based on A-weighted sound pressure levels would suggest otherwise.

REFERENCES

1. WHO. 2000. *Guidelines for community noise*. Geneva: World Health Organization.
2. WHO. 2009. *Night noise guidelines for Europe*. Copenhagen: World Health Organization Regional Office for Europe.

3. WHO. 2011. *Burden of disease from environmental noise*. Copenhagen: World Health Organization Regional Office for Europe.

4. ISO. 2003. *Acoustics—Assessment of noise annoyance by means of social and socio-acoustic surveys (ISO/TS 15666:2003(E))*. Geneva: International Standards Organization.

5. Berglund, B., A. Preis, and K. Rankin. 1990. Relationship between loudness and annoyance for ten community sounds. *Environmental International* 16: 523–531.

6. Watts, G., and P. M. Nelson. 1993. The relationship between vehicle noise measures and perceived noisiness. *Journal of Sound and Vibration* 164 (3): 425–444.

7. Nilsson, M. E. 2007. A-weighted sound pressure level as an indicator of short-term loudness or annoyance of road-traffic sound. *Journal of Sound and Vibration* 302(1-2): 197–207.

8. Zwicker, E., and B. Scharf. 1965. A model for loudness summation. *Psychological Review* 72(1): 3–26.

9. ISO. 1975. *Acoustics—Method for calculating loudness level (ISO 532-1975 (E))*. Geneva: International Standards Organization.

10. Nilsson, M. E., M. Andéhn, and P. Leśna. 2008. Evaluating roadside noise barriers using an annoyance-reduction criterion. *Journal of the Acoustical Society of America* 124(6): 3561–3567.

11. Rådsten-Ekman, M., B. Vincent, C. Anselme, A. Mandon, R. Rohr, J. Defrance, D. Van Maercke, D. Botteldooren, and M. E. Nilsson. 2011. Case-study evaluation of a low and vegetated noise barrier in an urban public space. Paper presented at the *Proceedings of Inter-Noise 2011*. INCE, September 4–7, Osaka, Japan.

12. Attenborough, K., I. Bashir, T. J. Hill, S. Taherzadeh, J. Defrance, and P. Jean. Noise reduction using surface roughness. In *Environmental methods for transport noise reduction*, eds. M. E. Nilsson, R. Klæboe, and J. Bengtsson (pp. 121–154). Oxford, U.K.: Spon Press.

13. Laszlo, H. E., E. S. McRobie, S. A. Stansfeld, and A. L. Hansell. 2012. Annoyance and other reaction measures to changes in noise exposure—A review. *Science of the Total Environment* 435–436: 551–562.

14. Brown, A. L., and A. Muhar. 2004. An approach to the acoustic design of outdoor space. *Journal of Environmental Planning and Management* 47(6): 827–842.

15. Brown, A. L. 2012. A review of progress in soundscapes and an approach to soundscape planning. *International Journal of Acoustics and Vibration* 17(2): 73–81.

16. Axelsson, Ö., M. E. Nilsson, and B. Berglund. 2010. A principal components model of soundscape perception. *Journal of the Acoustical Society of America* 128: 2836–2846.

17. Lavandier, C., and B. Defréville. 2006. The contribution of sound source characteristics in the assessment of urban soundscapes. *Acta Acustica United with Acustica* 92: 912–921.

18. De Coensel, B., D. Botteldooren, T. De Muer, B. Berglund, M. E. Nilsson, and P. Lercher. 2009. A model for the perception of environmental sound based on notice-events. *Journal of the Acoustical Society of America* 126: 656–665.

19. Moore, B. C. J. 1995. Frequency analysis and masking. In *Hearing*, ed. B. C. J. Moore (pp. 161–205). London: Academic Press.

20. Watson, C. S. 2005. Some comments on informational masking. *Acta Acustica United with Acustica* 91: 502–512.
21. Lutfi, R. A., L. Gilbertson, I. Heo, A.-C. Chang, and J. Stamas. 2013. The information-divergence hypothesis of informational masking. *Journal of the Acoustical Society of America* 134(3): 2160–2170.
22. Watts, G. R., K. V. Horoshenkov, and L. Ragonesi. 2009. Measurement and subjective assessment of water generated sounds. *Acta Acustica United with Acustica* 95: 1032–1039.
23. Nilsson, M. E., J. Alvarsson, M. Rådsten-Ekman, and K. Bolin. 2010. Auditory masking of wanted and unwanted sounds in a city park. *Noise Control Engineering Journal* 58(5): 524–531.
24. Galbrun, L., and T. T. Ali. 2013. Acoustical and perceptual assessment of water sounds and their use over road traffic noise. *Journal of the Acoustical Society of America* 133: 227–237.
25. Oldoni, D., B. De Coensel, M. Boes, M. Rademaker, V. De Baets, T. Van Renterghem, and D. Botteldooren. 2013. A computational model of auditory attention for use in soundscape research. *Journal of the Acoustical Society of America* 134: 852–861.
26. Rådsten-Ekman, M., Ö. Axelsson, and M. E. Nilsson. 2013. Effects of sounds from water on perception of acoustic environments dominated by road traffic noise. *Acta Acustica United with Acustica* 99: 218–225.
27. Hong, J. Y., and J. Y. Jeon. 2013. Designing sound and visual components for enhancement of urban soundscapes. *The Journal of the Acoustical Society of America* 134(3): 2026–2036.
28. De Coensel, B., S. Vanwetswinkel, and D. Botteldooren. 2011. Effects of natural sounds on the perception of road traffic noise. *Journal of the Acoustical Society of America* 129(4): EL148–EL153.
29. Culling, J. F., and M. A. Akeroyd. 2010. Spatial hearing. In *The Oxford Handbook of Auditory Science: Hearing*, ed. C. J. Plack. Oxford, U.K.: Oxford University Press.
30. Bolin, K., A. Kedhammar, and M. E. Nilsson. 2012. The influence of ambient sounds on loudness and annoyance of wind turbine noise. *Acta Acustica United with Acustica* 14: 741–748.
31. Maffei, L., M. Masullo, F. Aletta, and M. Di Gabriele. 2013. The influence of visual characteristics of barriers on railway noise perception. *Science of the Total Environment* 445–446: 41–47.
32. Pheasant, R. J., G. R. Watts, and K. V. Horoshenkov. 2009. Validation of a tranquillity rating prediction tool. *Acta Acustica United with Acustica* 95: 1024–1031.
33. Pheasant, R. J., M. N. Fisher, G. R. Watts, D. J. Whitaker, and K. V. Horoshenkov. 2010. The importance of auditory-visual interaction in the construction of 'tranquil' space. *Journal of Environmental Psychology* 30: 501–509.
34. Jeon, J. Y., J. Y. Hong, and P. J. Lee. 2011. Soundwalk approach to identify urban soundscapes individually. *The Journal of the Acoustical Society of America* 134(3): 803–812.
35. Hong, J. Y., and J. Y. Jeon. 2013. The effects of audio–visual factors on perceptions of environmental noise barrier performance. *Landscape and Urban Planning* 125: 28–37.

Chapter 10

Economic analyses of surface treatments, tree belts, green façades, barriers, and roofs

Ronny Klæboe and Knut Veisten

Abstract. Environmental noise-control measures developed or refined through the HOSANNA project have the potential of being cost efficient. In urban contexts similar to the situations analysed, they are expected to provide more benefits to society than they cost to implement. The first reason for their good economic performance is that environmental measures like lattice barriers or tree belts can be relatively inexpensive to establish. The second is that, in the cases involving dual-layer porous asphalt with and without resonators, the increased acoustic performance more than compensates for the additional costs. The third reason is that measures, such as vegetated façades, provide additional nonacoustic amenity and aesthetic benefits. Some measures, such as tree belts, are both inexpensive and provide such additional benefits.

10.1 INTRODUCTION

10.1.1 Societal cost-benefit analyses in HOSANNA

In this chapter, we present economic analyses of green roofs, green façades, and surface treatments. We also analyse the impact of combining at source and propagation measures to shield a community; more precisely, the use of dual-layer porous asphalt surfaces with (and without) resonators in addition to brick lattices. Moreover, we present economic analyses of a low vegetative barrier. Finally, we do a comprehensive set of economic calculations of tree belts, with and without artificial elements.

The economic analyses indicate that several noise-control measures based on the use of urban greenery have the potential of being cost efficient (provide more benefits than costs). In several cases, they are robustly efficient — providing benefits of more than twice the cost.

The best performing environmental measures are relatively inexpensive to establish (lattice barriers, tree belts), and/or provide high acoustic benefits (porous asphalt with and without resonators), and/or generate amenity/aesthetic benefits (vegetated façades) in addition to their noise-reducing effects (Figure 10.1).

221

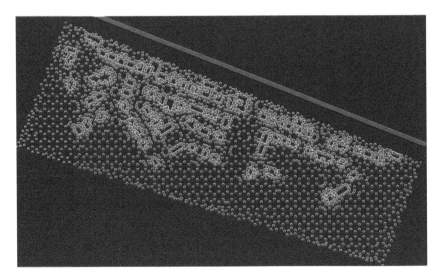

Figure 10.1 Residents of a community, showing noise reduction calculation points for 73 buildings and ground.

As part of the work to assess the environmental measures potential for improving the life quality of European citizens, we have made a first attempt to establish unit values for the aesthetic/amenity benefits of green measures. When including all relevant effects of a noise-control measure— as we should when addressing the economic rationale of implementing a measure—some measures become cost efficient that, from a purely acoustic perspective, are not worth the investment.

Confidence intervals that inform on the uncertainty of a given estimate are mandatory in many research settings providing input to important decision making. We argue that this also should be a requirement for economic indicators used to promote or argue against environmental projects affecting the life quality of the citizens of Europe. We have subsequently integrated a facility for undertaking Monte Carlo simulations to the open access web framework developed for undertaking the economic analyses.

Cost–benefit analyses are made *ex ante*, (beforehand), which is the typical situation for input to decision making. However, also the effects of the measures are estimated *ex ante*, often based on simulations. The sizes of the calculated effects need to be verified through field studies.

We use a subscript, for example $€_{2002}$, to show the year (here 2002) a monetary value was obtained. Prices and valuations are all updated to the project year by means of the harmonized consumer price indices for EC27 (yearly averages).

10.1.2 Economic analyses of six groups of measures

We present results from cost–benefit analyses of six groups of measures. The main results and the uncertainty associated with the results are presented through the overviews provided from the Monte Carlo simulations that are part of a web-based Open CBA (cost–benefit analysis)[1] programme used for the calculations.

For the analyses, we have adopted the EC (European Community)-wide HEATCO recommendations of an annual discount rate of 3% and a project horizon of 40 years.[2] With the above settings, the annuity factor is 23.8082. To cover an up-front investment of €$_{2011}$ 1,000 over a project period of 40 years, one needs to pay a yearly amount of not €$_{2011}$ 1,000/40 = €$_{2011}$ 25, but €$_{2011}$ 1,000/23.81 = 42.00 or 68% more than if the discount rate had been zero. We use the valuation result for noise reduction impact of €$_{2002}$ 10.095 per dB(A) per person and per year as derived from Bickel et al.,[2] as have other authors; for example, Andersson et al.[3] After the HOSANNA project ended, there has arisen some uncertainty as to whether the valuations should be scaled downwards (see Bickel et al., Table 6.11[2]). However, as this issue is not fully clarified, we have retained the Euro values per dB(A) per exposed. They also are similar to other recommended noise valuations for Europe (see Nijland,[4] Figure 3.2), or possibly on the lower side (see Andersson et al.[3]).

We use market, not factor prices. If the provision of environment benefits is seen as an alternative to producing other goods, prices before tax (factor prices) could be used instead.

From valuation studies, we have estimated that the aesthetic/amenity benefits of green roofs/façades are €$_{2010}$ 2.42 per square metre per resident per year and the presence of trees in the front or back of a house €$_{2010}$ 0.5 per square metre canopy per person per year.[5] We use the HEATCO valuation of CO_2 for the period 2010–2019 of €$_{2002}$ 26/tonne.

The six groups of measures include:

1. Vegetated roofs and rooftop vegetated barriers applied to a residential complex.
2. Vegetated façades. Here we introduce aesthetic/amenity effects in addition to the acoustic benefits. The measures are applied to a residential complex with 48 apartments and 115 residents.
3. The third group of measures are surface treatments in the form of brick lattices that provide noise benefits without being expensive. These are applied between a four-lane main road and a residential community described in the QUIESST (QUIetening the Environment for a Sustainable Surface Transport) project.
4. The fourth group presents a low vegetated barrier in the same residential situation, but the source here is dual tram lines. Follow-up studies of aesthetic/amenity benefits are motivated.

5. The fifth group of economic analyses deals with porous asphalt with and without a subsurface lattice of resonators. The resonators are in the form of solid bricks containing hollow openings that function as Helmholz resonators. The bricks are totally covered with asphalt. Situation analysed is the same as for the third group of measures analysed.

6. The sixth and last group of economic analyses is of the use of tree belts. Different alternatives where a 200-m tree belt with and without artificial components is inserted between a four-lane main road and a resident community are used.

10.1.3 Benefit-cost ratios applied for ranking projects within groups

We have used the benefit–cost ratio (BCR) output from the economic cost–benefit analyses (CBA) as a simple means of categorizing and ranking projects.

For each group of measures, we provide the summary overview illustration that the Monte Carlo package provides. Here stacked bar charts, one for each measure, are aligned vertically to allow easy comparison of the results (Figure 10.2).

The concept of the BCR is deceptively simple. By assigning monetary values to all benefits reaped from a project, the sum benefits can be calculated. If accumulated benefits exceed accumulated costs, we end up with a BCR greater than one and the measure is deemed *cost efficient*. As estimates of costs and benefits are uncertain, one might require that a measure is estimated to be *robustly efficient*; that is, providing benefits over the lifetime of the project that exceeds twice the cost. Furthermore, if several cost-efficient projects compete for public funding, the project with highest BCR is normally the best candidate for funding.

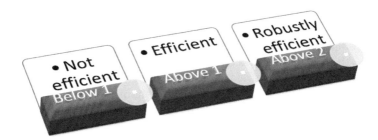

Figure 10.2 The size of the BCR allows us to sort measures in three categories: "less than 1," "greater than 1," and "greater than 2."

10.1.4 It is really projects, not measures, that are assessed economically

One should keep in mind that although the analyses focus on different noise abatement devices or treatments, the economic analysis is an analysis of improvement projects within a given context. Cross-comparisons of BCRs between groups of measures need to assess how the changes in context between different groups of measures affect the results.

Even within the same group of measures, the economic performance depends critically on how many people live close enough to benefit from the measure, as well as favourable or unfavourable implementation conditions etc. Location, land prices, the availability of materials, local prices, and contextual factors all affect results.

A closer description of the settings for the economic analyses and Monte Carlo simulations are given in Section 10.8.5. The derivation of unit values for the value of aesthetic/amenity effects associated with the presence of trees in front or in the back of a house and the aesthetic/amenity effects of green façades and rooftops into the analyses are described in Sections 10.9.4 and 10.9.6.

10.2 ECONOMIC ANALYSES OF GREEN ROOFS AND ROOF BARRIERS

10.2.1 Extensive roofs, roof barriers, and surface treatment alternatives

We have undertaken economic analyses of some of the rooftop treatments described by Kang et al. in Chapter 8. Most of these are extensive roof treatments—relatively inexpensive solutions where the roof is covered with simple sedum plants or other plant species that are resistant to drought. A couple of the alternatives were rooftop barriers—small vegetated barriers along the rim of the roof. The investment and maintenance cost of these are higher than for the extensive roof treatments (Table 10.1).

One of the roof surfaces analysed did not consist of any vegetation at all, but instead was a surface treatment in the form of a periodic array of inverted U-formed tiles.

10.2.2 Input to the economic analyses

The economic analyses were undertaken for a standard building block situation consisting of a six-storey building with 48 apartments. Given an average household size of 2.4,[6] and multiplying, we estimated the number of beneficiaries (residents of the 48 apartments) to be 115. All input factors were provided in a tabular form (Table 10.1). The acoustic input factors are

Table 10.1 Input factors for the economic analyses of vegetated roofs and one roof covered with materials introducing a periodic roughness

	Benef	Base dB(A)	Δ dB(A)	m²	Invest cost €	Maint year €	Life span
Vegetated flat roof	115	51.4	2.7	688	40	2	50
Vegetated shed roof, angled street canyon	115	51.4	7.0	688	40	2	50
Vegetated gabled roof	115	51.4	8.2	688	40	2	50
Vegetated roof with one convex corner	115	51.4	6.5	344	40	2	50
Vegetated roof with three convex corners	115	51.4	6.5	492	40	2	50
Vegetated roof with one concave corner	115	51.4	2.7	688	40	2	50
Vegetated roof with one corner sloped	115	51.4	3.7	688	40	2	50
Periodic roof roughness, coarse	115	51.4	1.9	737	40	2	50
Rooftop barrier canyon courtyard	115	51.4	2.8	173	500	25	10
Rooftop barrier 0,64 × 0.96	115	51.4	0.8	115	500	25	10

Note: Benef = number of residents; Base dB(A) = average exposure before noise reduction; Δ dB(A) = noise reduction; m² = treated area; Invest cost € = investment costs per m² and year; Maint € = maintenance costs per m² and year; Life span = expected life span (years).

based on individual calculations for each of the resident locations, but are here given as changes in the mean exposure.

For the extensive roof surfaces, the life span was estimated to be 50 years given a yearly maintenance expenditure of 5% ($€_{2011}$ 2) of the investment cost ($€_{2011}$ 40). For the intensive alternatives (rooftop barriers), the investments are larger ($€_{2011}$ 500), and the yearly maintenance is kept at 5%, which is $€_{2011}$ 25. (Because the life span of the intensive vegetation alternatives is much shorter, the relative size of the investment costs increases.) It was assumed that none of the proposed green roofs would provide any aesthetic benefit.

Some of the noise reductions resulted in noise levels below our chosen cutoff of 45 dB(A) (see Table 10.1, second row, for an example). Here, the noise levels are reduced to 44.4 dB(A). However, in these cases, the last (0.6) dB(A) noise reductions are not considered to bestow any additional benefits on the residents that are affected. The main reason is that at such low noise levels other noise sources "take over" and dominate. An additional reason is that we already use 45 dB(A) as a cutoff value or 5 dB(A) less than 50 dB(A) given in the HEATCO-study.[2] To reduce the lower limit further would imply extrapolating valuation results to noise situations that are qualitatively different from those used to obtain the monetary values.

Because the acoustic improvement affects only the courtyard, the impact with respect to annoyance reduction is assumed to be significantly lower than if the exposure on the most exposed façade had been reduced. Based on our own research and international results,[7–14] we have tentatively assessed that the size of the effect is 30% that of reducing the noise level on the most exposed façade. This is a slightly higher conversion factor than the 20% proposed by the CityHush project.[15] We subsequently count a 3-dB(A) reduction on the quiet side as achieving the equivalent of 0.9 dB(A) on the most exposed façade.

10.2.3 Three measures are cost efficient, one of which is robustly efficient

Using the Monte Carlo facility of the Open CBA programme,[1] it is easy to compare the economic performance of the different alternatives and add a complete sensitivity analysis of the effect of changes in input values.[10]

In Figure 10.3, the stacked bars show the median BCRs for each of the various alternatives as a black vertical line. The percentiles (5, 20, 35, 50, 65, 80, 95) are shown with the range 35%–50%–65% in dark blue.

A lighter yellow background is used to highlight measures that are *cost efficient*. All measures where the average result (vertical black line) lies to

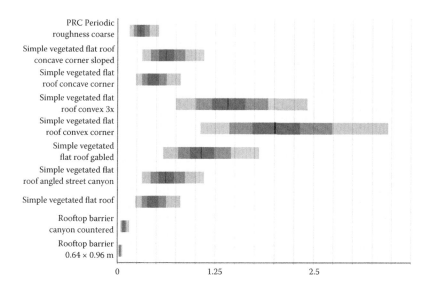

Figure 10.3 Monte Carlo simulation results; vegetated roofs and rooftop barrier.

the left are *not efficient*. Here the costs exceed the benefits. Measures where the median values lie to the right of the yellow area are *robustly efficient*.

The major determinant for the variation in the BCRs of the various roof alternatives is the size of the noise reduction versus the size and, consequently, the cost of the treatment/measure (Table 10.1). From the overview (Figure 10.3), it is clear that only one of the measures is robustly efficient (simple vegetated flat roof with convex corner), and an additional two measures cost efficient (simple vegetated flat roof gabled and simple vegetated flat roof with three corners).

The reason the simple vegetated flat roof comes out on top is that a good acoustic effect is achieved from treating a relatively smaller area saving both investment and maintenance costs, and that the base line noise level is high enough that almost the whole reduction provides real benefits (is not affected by the cutoff).

Rooftop barriers have a short life span, are relatively expensive, and, given that they fail to provide aesthetic benefits, seem not to make good economic sense.

10.3 ECONOMIC ANALYSES OF VEGETATED FAÇADES

10.3.1 Input to the economic analyses

Input factors for the CBA are shown in Table 10.2. The columns are, from the left, the number of beneficiaries (residents that receive an acoustic and aesthetic improvement), the premeasure baseline noise level, the calculated average quiet side improvement in dB, the number of square metre "greenery," the investment cost per m², and the maintenance cost per m². The (rightmost) column contains the life span of the vegetated façade — the time interval before it needs to be replaced. Extending the life length of a device/treatment can have a profound effect on the economic viability of a measure. Not only is the average cost per year reduced, but the stream of reinvestments are pushed farther into the future, and discounted more heavily.

To illustrate CBA in more detail, we focus for pedagogical reasons initially on only two of the vegetated façade alternatives. The purpose is to make clear the impact of taking all relevant effects into account — not only the acoustic benefits.

10.3.2 Economic analyses of two vegetated façade openings

The first alternative is a 3-m–high vegetated façade opening, while the second alternative is a 19.2-m-high vegetated façade opening (Figure 10.4).

Table 10.2 Input factors for the economic analyses

	Benef	Base dB(A)	Δ dB(A)	m^2	Inv cost €	Maint €	Life span
Vegetated façades in canyon	115	51.4	1.0	2949	500	25	10
Vegetated façades upper half of canyon	115	51.4	1.0	1475	500	25	10
3-m-tall façade opening to cross-street	115	47.9	3.4	58	500	25	10
19.2-m-tall façade opening to cross-street	115	59.3	4.1	369	500	25	10
3-m-tall façade opening to street	115	56.3	4.5	58	500	25	10
19.2-m-tall façade opening to street	115	60.0	4.3	369	500	25	10

Note: Benef = number of residents; Base dB(A) = average exposure before noise reduction; Δ dB(A) = noise reduction; m^2 = treated area; Inv cost € = investment costs per m^2 and year; Maint € = maintenance costs per m^2 and year; Life span = expected life span (years).

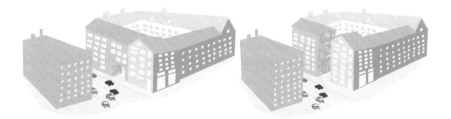

Figure 10.4 CBA analyses of two vegetated façade openings. (Illustration courtesy of Tove Hennix.)

Both sides of the façade openings are covered with vegetation. The treatment absorbs sound and reduces the amount of noise that enters through the opening in the façade and into the courtyard affecting the silent side of a 48-unit apartment complex. We show one set of results where we ignore aesthetics, only taking acoustic benefits into account, and, in a second set of results, the aesthetic benefits are included (Figure 10.5). (A few residents living in the neighboring building directly opposite the green walls also benefit aesthetically from the façade treatment.)

The total area covered by the façade improvements is (2 × 3 × 9.6) = 58 m^2 for the 3-m alternative and (2 × 19.2 × 9.6) = 369 m^2 for the 19.2-m alternative. The cost of the vegetated façades is set to €$_{2010}$ 500/m^2, with a lifetime of 10 years. This yields an equivalent annual

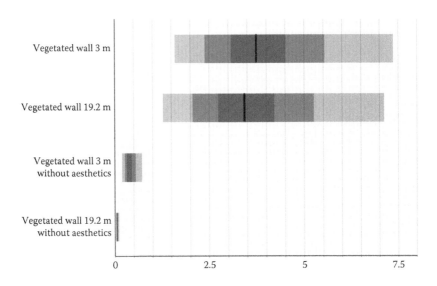

Figure 10.5 Vegetated façade opening projects analysed in the HOSANNA brochure.

cost of €$_{2011}$ 56.91/m². The annual maintenance cost is €$_{2011}$ 25/m². The 19.2-m-high green walls reduce the mean noise level by 4.1 dB(A), while the 3-m-high green walls reduce it by 4.5 dB(A). In all, 115 people benefit from the acoustic and aesthetic/amenity improvements. Because the acoustic improvement affects only the courtyard, the impact with respect to annoyance reduction is assumed to be significantly lower than if the exposure on the most exposed façade had been reduced. We have estimated that the effect is 30% that of reducing the noise level on the most exposed façade.[10] We subsequently count a 3-dB(A) reduction on the quiet side as achieving the equivalent of a 1-dB(A) reduction (to be exact, 0.9 dB(A)).

When taking aesthetic benefits into account, both noise abatement alternatives prove to be robustly efficient, providing benefits four times greater than the costs (BCR > 4). The aesthetic benefits, which are proportional to the costs, dominate the calculations and the two alternatives, consequently, prove to be about equally efficient.

Even allowing for substantial uncertainty in all estimated costs and benefits, the entire distribution of these BCRs is in the area of robust efficiency when aesthetic/amenity benefits are included.

When we ignore aesthetic benefits, neither of the alternatives is cost efficient, as both cost more than the benefits they produce and the BCR is less than one (BCR < 1). Since the 3-m-high vegetated façade alternative provides a somewhat higher noise reduction for a substantially lower investment, it is about six times more cost effective than is the 19.2-m alternative.

From the Monte Carlo simulations, there is virtually no chance that the acoustic benefits alone will make the measures cost efficient. When we use the information provided on the input factors and their uncertainties, the measures are almost certain to cost more than they are worth in terms of acoustic improvement alone.

Here, the benefits from the aesthetics outweigh by far the acoustic benefits. This could be part of the explanation for why architects and urban dwellers pay so much attention to the visual aesthetics of areas and why many building projects try to attract customers by making parks, quieter areas, green roofs, and façades an integral part of their offerings to the public.

Aesthetic benefits will consequently be included in all CBAs where applicable. However, we want to stress that the estimated unit value applied for aesthetic/amenity values of green roofs and façades is a weighted average of only a relatively small number of fairly heterogeneous studies.[10]

10.3.3 All vegetated facade projects are robustly efficient when aesthetic appreciation is included

The Monte Carlo results (Figure 10.6) show that all vegetated façades are robustly efficient when aesthetic appreciation of the vegetated walls is included in the CBA. From the spread of results, we also see that it is very seldom (less than 5% of the cases) when a green wall is not economically efficient, given the applied unit value.

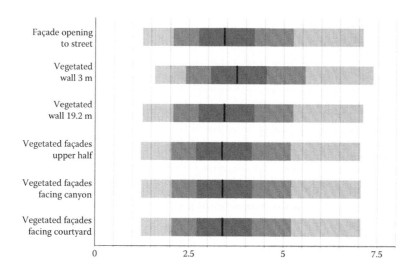

Figure 10.6 Monte Carlo simulations for vegetated façades; aesthetics included in CBA.

The Monte Carlo simulations of the BCRs taking aesthetics into account show that the results are relatively similar even though the project sizes differ. This is a result of how the BCR is constructed. The aesthetic benefit is more or less proportional to the area covered, as is the cost of constructing and maintaining the vegetated façades. It follows that the ratio between the benefits and the costs stays more or less the same, even though total costs of alternatives requiring hundreds of square metres are compared against smaller projects, such as covering a 3-m façade entrance with vegetation on both sides.

10.4 ECONOMIC ANALYSES OF SURFACE TREATMENTS

10.4.1 Lattices with and without maintenance

Economic analyses have been undertaken of hard surface treatments.[16–18] The surface treatments studied are in the form of brick lattices. Two 600-m-long lattices are considered—one placed on the central reservation between the opposing lanes, and one at the road side located between the road and a residential community and the four-lane road. The brick lattices reduce the noise exposure for residents of 74 two-storey-high buildings (Figure 10.7).

10.4.2 Cost calculations

We have included land usage costs, clearance costs, construction costs, and, for one of the alternatives, maintenance costs. From the imported files and the resulting graphical representation of the source and receiver positions, we have assessed the length of the lattices along the central reservation and the near roadside to be about 600 m in length. Each lattice covers 1,800 m^2.

10.4.3 Land usage costs

When the land areas along the road are dedicated to serve the adjacent space needs of the road (as often will be the case), there is no viable alternative usage, and the cost is zero. This should be true for any lattice or device located on or along the central reservation separating opposing traffic.

For the lattice located on the proximal side of the road, we have applied an average agricultural land price ($£_{2012}$ 15,500 per hectare), giving a unit price of $€_{2012}$ 1.9 per square metre. Because the price of land in the outskirts of a city area will be higher, we use a triangular distribution in the Monte

Figure 10.7 Lattice prototype.

Carlo simulations to allow the right tail to extend to 10 times this amount. Note that socioeconomic calculations differ from private deliberations in that we do not count ownership transfers or the fact that money is spent to purchase land as a "cost." However, not being able to put the land to other good use, e.g., biomass production, counts as a loss.

10.4.4 Clearance, construction, and maintenance costs

We apply an average of €1.5 per square metre. This is the cost of preparing the ground for the deployment of prefabricated modules or on-site constructed measures.

Establishing the brick lattices consists of materials, work, and equipment costs. A 100-m-long, 0.3-m-high, 3-m-wide lattice array (total area 300 m^2) is specified to be requiring the equivalent of $(1{,}670 + 250) \times 2 = 3{,}440$ bricks, costing £3.02 per brick. This results in a total cost of materials of £10,390. Labour and plant hire are estimated to cost £1,000/day for 10 days (i.e., £10,000), which is in addition. This gives a total cost per 300 m^2 of £$_{2011}$ 20,390 or a unit price of £$_{2011}$ 68/m^2.

Converted to Euros, we obtain a unit price of €$_{2011}$ 78.46/m^2. Because the lattice along the central reservation also serves residents on the other side of the road, we count half the cost. (If we arrive at a situation where residents' locations on the other side of the road were mirror images of the layout positions described in Figure 4.2, each of the resident groups would need to cover the cost of the lattice on their side of the road, plus half the cost of the lattice on the central reservation.)

We are simplifying the calculation by excluding both benefits and costs of the residents on the other side. We, therefore, exclude the cost of the lattice on their side from the calculations. It follows that we also should exclude their half of the cost of the lattice on the central reservation (because we have three lattice structures—one on each side and one in the middle—and two communities on each side of the road, it means that each group needs to cover the cost of 1½ of the lattice structure).

To cover the situation where there are only dwellings on one side, and there is nobody with whom to share the cost of the lattice on the central reservation, we have done the economic calculations for a scenario where benefits of the residents on one side of the road must bear the full cost of both the central and road-side lattice.

The lattice has a specified life span of 20 years. For the economic calculations, this seemed a bit conservative considering that, in production, the prototype would be replaced by prefabricated stable and interlocking modules. Consequently, we have also analysed the economic performance of an alternative deployment regime where the life span of the lattices is extended to 40 years by adding in a maintenance effort where 10% of the bricks/modules are replaced every decade. (The life span is still considerably shorter than the minimum 60-year durability requirement for standard concrete noise abatement walls.)

10.4.5 Noise reduction benefit calculations

Based on high-quality insertion loss calculations for the lattice itself, an engineering model[19] was utilised to calculate L$_{den}$ exposure values for calculation points on the ground and on the façades of 74 two-storey-high buildings (see Figure 10.8). For extracting the information from these calculations and that which was necessary for the CBA, a data interchange

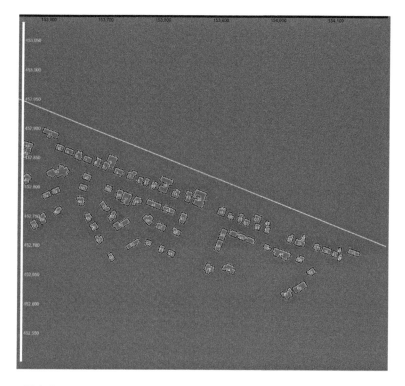

Figure 10.8 Building outlines with calculation point locations. One lattice is placed on the central reservation between opposing lanes, whereas the other intervenes between the 2 × 2 lane road (diagonal line at the top) and the 6-m-high, two-storey buildings.

format was specified. In all three files containing information on the source (four-lane road), the context (buildings) and calculation points on the ground and on the buildings were processed.

The geographic information was converted to a text-based information interchange format (JSON) and stored together with other information in the OCBA (open cost benefit analysis) system. Display functions were constructed to portray the information on a computer screen in a scalable vector graphics format (.svg) amenable for visual inspection.

To undertake CBA, we need an adequate estimate of the noise reduction impact for each resident. Information on the noise exposure changes at each of the numerous calculation points were post-processed to find the equivalent noise reduction impacts for the resident population.

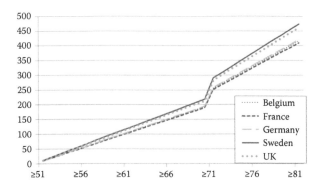

Figure 10.9 Valuation of noise exposure reductions in the HEATCO study[2].

10.4.6 The relationship between noise reduction and "kverks"*

In the HEATCO study,[2] high noise exposure values ($L_{den} > 70$ dB(A)) are thought to have more serious health effects in addition to their effects on annoyance, sleep disturbances, and life quality (Figure 10.9). The value of decreasing noise by 1 dB in this high-exposure region is assessed to be double that of reducing noise exposure by 1 dB(A) in the interval 50 dB(A) to 70 dB(A). (In our work, we use a cut-off value of 45 dB(A) instead of 50 dB(A).) In addition to the steeper slope, there is a bonus when transitioning from the high exposure range to the lower.

Monetary valuation experiments and hedonic pricing studies assign a monetary value not to an exposure reduction, but to the impact of exposure reduction. Because there is no simple relationship between the exposure reduction in dB(A) and the size of the impact, we measure the impact in "kverks" rather than physical exposure units (dB(A)). Making such a distinction was necessary in the economic analyses of green roofs and also on vegetation where the impact of noise reduction on the quiet side was deemed equivalent to a 30% impact of the noise reduction on the most exposed side.) We define one kverk as a one-year improvement in life quality associated with a noise reduction (L_{den}) of 1 dB(A) on the most exposed façade for a person exposed to between 46 and 70 dB(A). The conversion factor between exposure and impact for one resident and the impact in kverks is shown in Table 10.3.

In addition to the conversion factors, a one-time bonus equivalent in size of three kverks is given when noise exposure is no longer above 71 dB(A). This is the region where noise exposure is thought to have serious health effects in addition to noise annoyance, sleep deprivation, and other disturbances.

---

* 1 kverk is the impact of a 1 dBA change in noise exposure in the interval 46–70 db(A).

Table 10.3 Relationship between a 1- dB(A) noise reduction affecting one resident and its impact in kverks

dB(A)	Context	Kverks
0–45	All	0.0
45–70	Quiet side	0.3
45–70	Most exposed	1.0
71–	Most exposed	2.0

Note: A 1 dB(A) change in exposure of the noise source in focus gives an impact change of 1.0 kverk under ordinary conditions. Depending on the baseline noise level and context, it could be higher, or zero.

As a measure of impact, kverks follow simple rules of addition. Two persons having an impact of two kverks count as a total impact of 4 dB(A).

10.4.7 Noise reduction impacts for residents in 74 buildings

For each calculation point, a base value (the exposure in the zero-situation without any abatement) and the exposure after the alternative lattice arrangements were calculated. Because the different buildings affected by traffic noise had a different density of calculation points, we could not use the grand mean of all building calculation points to estimate noise reductions for the residents. For several of the buildings, some of the exposure values were above 70 dB(A) or below the cutoff value of 45 dB(A), thus, simply averaging noise exposure changes would be misleading as well. We, therefore, designed algorithms to calculate the noise impact changes in kverks for each calculation point and averaged these for each building. In the next step, we used the building as characteristics for all residents of the respective buildings.

The number of apartments in each building was calculated by dividing the building surface area (ground surface area multiplied with the number of floors) by an estimated average apartment size of 70 m². The number of residents benefiting from the measure was calculated by multiplying the number of apartments with an average household size of 2.4. (Both the apartment size and number of persons per apartment vary in different countries.)

Given the nominal noise reductions, we have calculated the accumulated noise benefits equal to 1,649.37 kverks before counting in bonuses and maluses (good and bad). To account for the extra weight given to health-promoting noise reductions above 71 dB(A), we add in 123.9 kverks. We also add in 37.2 bonuses of 3 kverks from transitions out of the high exposure range. This brings the sum noise benefits to 1,885 kverks.

The total ground surface area (8,994) of the buildings multiplied by the two floors is 17,989 m². This means 617 residents receive an average noise benefit from the device(s) of 1,885/617 = 3.06 kverks.

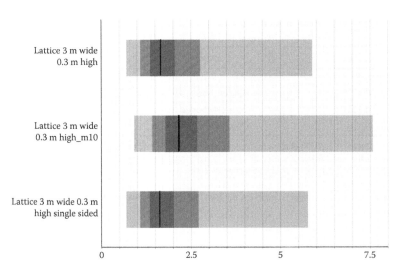

Figure 10.10 Results from the Monte Carlo simulations. Only in 15% of the simulation results do the measures fail to be efficient. The cost efficiency can be very good (long right tail).

10.4.8 Alternative with maintenance to prolong life span is robustly efficient

Given what we considered a somewhat conservative estimate of the life span of the brick lattices, we also have calculated a scheme where 10% of the bricks are replaced each 10th year, thereby extending their effective life span to 40 years. If the assumption of an expected lifespan holds, the economic performance of the measure becomes robustly efficient with benefits more than twice the cost.

The brick lattice without maintenance is cost efficient with good margin; however, it is not robustly efficient.

When we assess the costs and benefits of a lattice with maintenance in a single-sided exposure situation, the cost of the lattice on the central reservation is no longer shared, and the BCR is somewhat lower (Figure 10.10).

10.5 ECONOMIC ANALYSES OF LOW, VEGETATED BARRIERS

10.5.1 Prototype of low, vegetated barrier

The low barrier is a 1-m-high, 40-cm vegetated barrier. It is a commercial product delivered by Canevaflor.[20] It is here considered as a noise abatement measure shielding a community from two tramway lines. The same layout of the affected community is used as for the brick lattices.

Figure 10.11 Low barrier in Lyon, France.

Whereas the vegetated barrier is perceived as both acoustic and aesthetically pleasing, the roadside placement means that it is not reasonable to assign the same aesthetic value as vegetated façades and roofs in the immediate vicinity of the residents.

The results from a small study in Lyon, France (Figure 10.11), suggested that most people appreciated the barrier (see Chapter 9).

10.5.2 Maintenance costs dominate

The benefit from the low barrier will depend critically on what use residents make of areas close to the tramway. Because we know little of this, we present our main results without taking aesthetic benefits into account.

From the results of the economic calculation, it is evident that the 5% maintenance effort each year weighs heavier on the cost side than the up-front investment.

A metal frame with noncorrosive materials means that the investment need not be replaced during the project period (here we use an 80-year life span for the frame mainly consisting of noncorrosive metal plates). In our calculations, we take into account residual values. If a device or treatment has a remaining usable value at project end, we deduct this value from the cost of the device or treatment during the project period. In our examples, the final year of the project is not a "hard" limit, but a calculation limit chosen for comparison purposes. We use the discounted value of the complete "unused" part of the device or treatment at project end; i.e., we assume "continued" business. (In the OCBA system, it is possible instead to specify that the values should be discarded or that only a proportion of the residual value is salvageable.)

10.5.3 Valuation studies of aesthetics of low, green barriers are needed

We have chosen to provide an example of a CBA counting in 10% of the aesthetic value used for green roofs and façades. This calculation does not have the same status as the previous calculations. It is simply meant to illustrate what effect this could potentially have, and as a motivation for further empirical studies (Table 10.4).

From the results, we can see that if the barrier has an aesthetic value 10% of what was assigned to green walls and roofs, this would increase the economic viability of this measure. However, assigning an aesthetic value to the low barrier is something that should be substantiated by empirical studies of how these types of vegetative low barriers are appreciated by the resident population.

10.6 ECONOMIC ANALYSES OF SOURCE + PROPAGATION MEASURES

10.6.1 Types of configurations

Measures targeting the propagation paths are natural supplements to measures targeting noise at the source. We, therefore, have undertaken economic calculations of a scheme where porous asphalt with and without resonators is used in conjunction with the brick/concrete lattices described in Section 10.4 (Figure 10.12).

Table 10.4 Annual costs and benefits of 600 m-long, 1-m-high vegetated barrier; aesthetics tentatively included

Component	Q*M*C*B	Annuity	Cost	Benefits	ΣCosts	ΣBenefits	ΣΣCosts	ΣΣBenefits	B/C Ratio
Maintenance	30	431.20	12,935.90						
Investment	600	16.01	9,604.52						
Aesthetics	6,120	2.50		15,269.63					
Noise	612	37.34		22,852.12	22,540.42	38,121.75	22,540.42	38,121.75	169%

Note: Q*M*C*B = Number of unit measures (Q) × unit measure size (M) × number of unit components/effects in a measure (C) × the number of beneficiaries (B); Annuity = the yearly "rent" that accumulated over the project period is equal to one or more investments; B/C Ratio = benefit-cost ratio.

Figure 10.12 German A24 section with "buried resonators" during construction. The resonators were covered with two-layer, open porous asphalt.

We consider the following configurations:

1. Lattices alone (see Chapter 6)
2. Lattices + dual porous asphalt versus Standard Mastic Asphalt (SMA)
3. Porous asphalt with resonators versus dual porous asphalt
4. The whole "package": lattices + porous asphalt with resonators

10.6.2 Lattice in combination with two-layer, open porous road surfaces

We have applied the various measure combinations to the same community layout as when we dealt with brick lattices alone (Figure 10.13).

When brick lattices are used, the noise impact calculations indicate that the larger benefits are enjoyed at some distance from the road. This is mainly a result of the acoustic properties of the lattices. Adding dual porous asphalt with or without resonators results in lower emissions from the road, but does not change the differential effect of the brick lattices.

10.6.3 Absolute and marginal kverk to dB(A) ratios

To explore the relationship between noise exposure and the improvement in life quality and life quality improvements in kverks, we have used

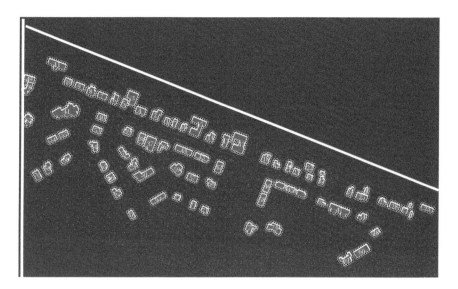

Figure 10.13 Community with 616 residents; brick lattice configurations.

the exposure calculations for the community depicted in Figure 10.13 (616 residents). As a departure point, we have the amount of kverks achieved with the lattice only (see Chapter 6). Because porous asphalt (with or without resonators) reduces the noise emissions from the road, we assume that the reduction in noise emissions translates in an equal reduction of the noise exposure values for each resident. Therefore, we can calculate the effect of reduced noise emissions from 1 dB(A) to 25 dB(A) and recalculate the kverk reduction obtained by the combined reduction in noise emissions and the reduction through the brick lattice. After dividing by the number of residents, we obtain a community average kverk response function that shows the relationship between reduced noise emissions and the impact on the community in kverks.

Given that the initial noise exposure from the four-lane road is quite high, one gets a larger benefit from the initial noise reduction in dB(A) than the following. As the noise abatement gets more effective, fewer and fewer residents are located in buildings exposed to noise levels above 71 dB(A). After some time, there are none left, so no more 200% kverk reductions or bonuses are obtained. At the lower end of the noise exposure range, the efficacy of further noise reductions diminishes due to the increasing number of residents below the cutoff limit of 45 dB(A) (Figure 10.14(a)(b)). This is an example of following the well-known economic law of diminishing returns.

After a 35-dB(A) reduction, there are no further reductions of the impacts, because we have reached the cutoff value for all residents.

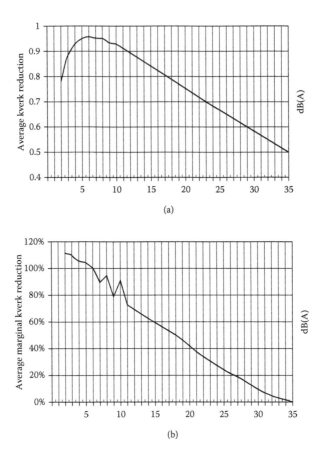

Figures 10.14 The average kverk reduction per dB(A) noise reduction: absolute and (b) marginal effect.

10.6.4 Special considerations when analysing combinations

Given that dual-layer, open-porous asphalt produces a noise benefit of 7 dB(A) on the average, we use a multiplier of 90% to take the diminished kverk reduction of the last dBs into account. When more than one measure is used to achieve an effect, the sequence in which they are applied matters. The first one applied will often produce a somewhat better benefit-cost ratio than the next. When dual-layer, open-porous asphalt is applied after the brick lattices, we are in the range of average combined noise reduction of 10 dB(A) and in the area of approximately 80% average efficacy.

Note that the economic analyses only take attribute changes into account. When two alternatives both feature a measure, such as a brick lattice, there is no difference in benefits and costs and, thus, they are omitted from all calculations. (We nevertheless have to take into account the indirect implications of the lattice in that the baseline and efficacy of further noise reductions are changed by the previous application of the lattice.)

10.6.5 Adding dual porous asphalt (with/without resonators) makes solutions robustly efficient

From the overview presented in Figure 10.15, we see that an economic analysis of adding resonators to dual porous asphalt indicates this is a good idea. The main reason is that, though the porous asphalt costs more, the resonators increase the life span of the road surfaces, which means that their lifetime cost is about the same as for dual porous asphalt without resonators. Hence, one gets the additional noise benefits for free.

The analyses also show that resurfacing the four-lane road makes good economic sense and most probably provides benefits more than twice the cost. Without resonators, there is a higher chance that it is merely cost efficient and not robustly so. Combining the lattice and dual porous asphalt with resonators, the entire BCR distribution is firmly in the area of robust efficiency.

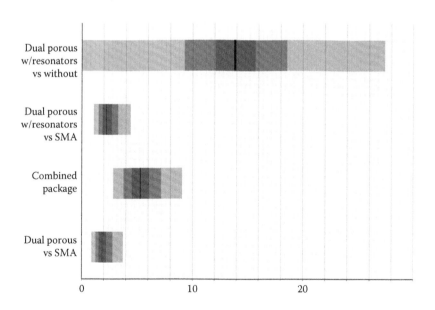

Figure 10.15 Monte Carlo simulations of combined brick lattices and porous asphalt with and without resonators; different sets of calculations.

10.7 ECONOMIC ANALYSES OF TREE BELTS

10.7.1 Tree belts used alone and in combination with artificial elements

One challenge with respect to using a tree belt as a noise abatement measure is that trees need space to thrive. Trees compete with their neighbours for water, nutrients, and sunlight. Based on expert opinion, trees planted in rows with 2 m between them and a distance between tree in each row of 1 m, which is a grid size of 1 × 2 m giving a density of 0.5 trees/m², should work well. This is what we have used for our economic analyses of tree belts (Figure 10.16). Planting trees closer could be detrimental to their well-being.

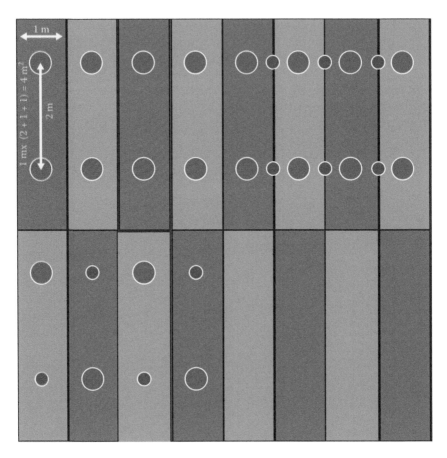

Figure 10.16 Layout with trees only, with poles inserted between trees in each row; sparse alternative where half of the trees are replaced with poles and soft ground.

If we don't limit ourselves to only using natural elements, we can augment the acoustic effect of a natural living tree belt by adding artificial noise abatement elements. We here explore two such alternatives. One is to add the equivalent of sturdy "garden poles" that are 2.5 m high and 12 cm in diameter. By inserting these impregnated wooden poles between the trees of each row, the gap between noise abatement elements (trees and poles) is narrowed down from 1 to ½ m. The other is to use more expensive "noise stoppers," quadratic or circular poles made from recycled waste products that are 25 cm in diameter, 1.5 m high filled with noise-absorbing material produced from scrap rayon.

10.7.2 Properties of artificial elements

Poles and noise stoppers produce their noise-reducing effect immediately, whereas one must wait some years before tree rods have grown sufficiently to give the same level of noise protection. Even though trees eventually provide most of the noise protection, the added artificial elements continue to give an extra noise reduction. By replacing defective and spent poles, the acoustic effect of the array of poles remains constant over the lifetime of the tree belt itself.

One could argue that wooden poles or other artificial elements will detract from the aesthetic value of the tree belt. However, newly planted trees are often supported by rods or different types of fences, especially if they bear fruit or it is important that the trees grow straight. We don't think wooden poles will be regarded as incongruent with the environment, or will reduce the aesthetic value of the tree belt. However, neither do they enhance the aesthetics of the tree belt.

10.7.3 Cost of tree belts

We use the assumption that the trees are pollarded (pruned). We use a land purchase price of €$_{2011}$ 188/m^2, which is the Lyon average.[21] To clear land in preparation for the tree belt, we use a cost of €$_{2011}$ 1.54/m^2. The cost per Salix tree rod 2 m high is given as €$_{2011}$ 0.55. Pollarding a tree provides 100 kg of wood cuttings, each fifth year starting from year six. The wood cuttings can be sold at €$_{2011}$ 30 per tonne. Planting (digging a hole and inserting plant) cost €4 per tree, and the cost of pollarding the trees (cutting and maintaining) cost €10 each fifth year starting from year six. The aesthetic benefits are calculated for half the number of residents, 272, half enjoying full 182 m^2 of tree crown on the average and 50% captured by a 0.75 multiplier. Each square metre canopy is valued at €$_{2012}$ 0.27/m^2 per person per year.

The residents enjoy an average of 9 dB(A) noise reduction from the tree belt when the reference case is rigid ground.

10.7.4 Cost of artificial elements

The cost of 2.5 m impregnated wooden poles from a Polish supplier is €$_{2008}$ 8.90. The life span is set at 12 years. Given a 100% × 100% configuration, there are 1,500 poles that need to be purchased. In addition to replacing the poles each 12th year, we have specified a maintenance activity where 2.5% of the poles are replaced each year.

In our explorations of how to use recycled materials to reduce noise, we have undertaken calculations of a novel product: "noise stoppers." These are oversized 25 × 25 cm rectangular hollow tubes with walls 2.5 cm thick made of recycled materials and filled with scrap rayon. The cost of a noise stopper pole is estimated to be €$_{2012}$ 20.80. We also here add in a maintenance activity where 2.5% of the noise stoppers are replaced each year. Based on the properties of the material, the life span of the noise stoppers is 40 years.

10.7.5 Tree belt 200 m long, 15 m wide to protect a community

To illustrate the change in life quality and health as measured in kverks, we have made use of a broader set of calculations undertaken by CSTB for a community of residents (Figure 10.17).

We have focussed on a 200-m-long tree belt. This is long enough that many residents are protected from noise coming in from the side. Because willows grow rapidly, we have been informed that the stem size can reach 60 cm. The speed of growth is high in the beginning. According to professionals dealing with willows for hedges, plants can have a diameter of 25 cm after only 10 year's time. A diameter of 12 cm (which is equal to the diameter of

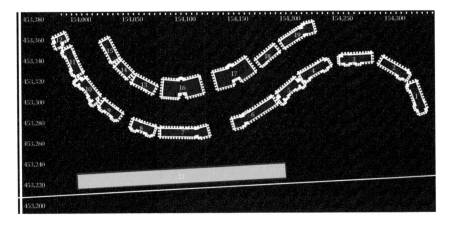

Figure 10.17 Community with 515 residents protected by 15-m-wide tree belt.

impregnated poles that are often used as support for garden fences) is reached after five years. However, it has been quite difficult to get solid information on growth characteristics with respect to stem size and we lack information on growth statistics for dense planting schemes. We have chosen to err on the conservative side. For average stem size, we use here a lower estimate of 17.5 cm as the average diameter during the life span of the tree belt.

10.7.6 Average kverk reduction as a function of stem size

The calculation has been repeated for tree belts with trees having stem sizes 4, 8, 12, 18, and 20 cm. This allows us to plot the average noise reduction in kverks as a function of stem size when placing a tree belt on a previously rigid ground (Figure 10.18).

We find the relationship to be linear from a stem diameter at breast height (dbh) from 4 cm to 20 cm. The equation is numerically simple: 6 kverks (the soft ground effect) + 0.3 dbh. This means that a 12-cm pole gives 0.3 × 12 = 3.6 kverks when assuming that the poles have the same frequency-independent absorption characteristics as the bark from trunks of living trees.

A "noise stopper" that is 25 cm in diameter should give at least 7.5 kverks. Because the structures are absorbing, they should reduce noise more than a tree stem does. The latter effect has not been included here. Tree trunks 20 cm in diameter should give 12 kverks including the forest floor effect. Trunks have an additive effect,[22] mainly because they affect a different part of the sound frequency spectrum. Additivity of the noise reduction by trees and poles is used here as well in a first approximation.

Whereas tree poles and noise stoppers have constant thickness, trees grow. Due to discounting, the benefits provided by the larger stem sizes

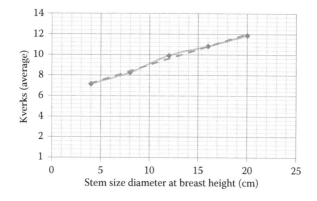

Figure 10.18 Community average kverk reduction as a function of stem size (diameter at breast height).

toward the end of the life span of the well-grown trees do not count as much as the benefits from the younger trees having smaller stem sizes. Therefore, we only add 3 dB to the 6 dB from the soft ground.

10.7.7 Amenity/aesthetic effect of the tree belt

Because we wish to employ the valuation that is given per square metre tree canopy,[5] we need to assess the canopy area that each resident benefits from, and then accumulate. We here assume that only the residents of the front-row houses benefit fully, and that those living on each side of the tree belt only benefit about 50%.

We assume that a resident benefits from trees within a 120-degree angle and 100 m distance. Given an average distance of some 30 m away from the tree belt, those who live closest to the tree belt benefit from about 70 front-row trees. If we assume that each additional row of trees adds half as much aesthetic and amenity value as the row before, approximately each resident benefits from the equivalent of two rows, or 140 trees. Given an average crown diameter of 1.5 m (radius 0.75) and 25% gaps in the coverage, each tree provides 1.3 m² canopy.

There will be seasonal changes in the aesthetic value of the trees. However, this variation is already incorporated when we use the hedonic pricing method. Changes in property prices reflect the average benefits over the full year.

10.7.8 Carbon sequestration

Trees convert carbon dioxide (CO_2) to carbon (C) and release oxygen (O_2) into the atmosphere. To find the amount of CO_2 that is sequestered by a tree, we need to find the amount of carbon that a tree contains. From this figure, we can calculate the amount of CO_2 that has been removed from the atmosphere by adding the molecular weight of two oxygen atoms to the molecular weight of carbon.

The trunk diameter at the end of the lifetime was set to 40 cm and the height at 2.5 m. Geometric considerations provide us with the aboveground volume and weight of the tree (cylinder, since the top is cut) ($2.5 \times 0.2 \times 0.2 \times \pi$). Nearly half the weight is simply water. We, thus, need to calculate the dry weight (multiply by ca. 60%). We also need to find the proportion of carbon in the dry tree. Here we assume the carbon content is half the dry weight (multiply by 50%). The root size is estimated to be 40% of the aboveground volume rather than 20% (multiply by 1.4). With a CO_2 price per tonne given by HEATCO to be €2002 26 per tonne, the benefit per year is limited, contributing 2% to total benefits from the tree belts.

10.7.9 Tree belt alternatives considered

We consider the tree belt configurations shown in Table 10.5.

Table 10.5 Tree belt configurations

Alternative	Description
Trees + poles 0.25	Half the trees are replaced by poles giving a density of 0.25 impregnated poles and 0.25 tree per m^2
Regular tree belt from scratch	Tree belt with density 0.5 tree per m^2 suboptimal regular planting scheme
Tree + poles 0.5 from scratch	Tree belt with density 0.5 tree per m^2 + 0.5 impregnated wood poles per m^2
Tree belt with "noise stoppers"	Tree belt with density 0.5 tree per m^2 + 0.5 noise stoppers per m^2
Tree belt: Baseline	Tree belt with density 0.5 tree per m^2 optimal planting scheme

10.7.10 Tree belt alternatives with artificial elements are best economic performers

From the overview of the BCR distributions, it is easy to compare the different alternatives. The last distribution is the analysis of a natural tree belt 15 m wide, 200 m in length on a 1 m × 2 m grid and with lifetime average stem diameter of 20 cm. It is robustly efficient. The aesthetic benefits are 20% of the total benefits.

However, adding poles or noise stoppers produce even higher BCRs. Each of the artificial additions results in about the same economic performance. The noise stoppers cost more, but have a longer life span and yield higher acoustic benefit (Figure 10.19).

Imposing strict grid regularity results is a significant reduction in the BCR. The same is true when we replace half the trees with poles so that we end up with a sparser grid.

10.8 ECONOMIC ANALYSES: SIMPLIFIED VERSION

10.8.1 The virtue of economic analysis: societal cost-benefit analysis

Investments in noise reduction measures are often guided by economic considerations. How do we choose between an acoustically efficient, but expensive device or treatment for reducing noise and an inexpensive, but less efficient method? When does the additional cost of adding extra noise abatement elements, the use of higher quality components, or increasing the size of a noise-reducing structure exceed the acoustic benefits? When is enough really enough? Both costs and effects are spread over a period of time. When does repeated yearly noise reduction benefits

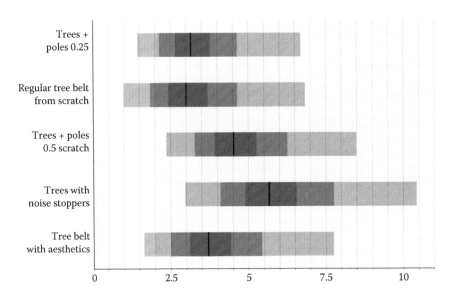

Figure 10.19 Monte Carlo simulations of various combinations of trees and artificial elements.

enjoyed over, say, 40 years outweigh a single, one-time expenditure in a noise-reducing treatment?

To answer such questions, we employ economic analyses.[23-26] These are "time machines," calculators that adjust for the timing of expenditures and benefits. By using a common project horizon to assess different abatement alternatives, a common discount rate, and methodology, the time factor is taken out of the equation. It then becomes possible to rank measures according to their overall economic performance. It makes it possible to compare measures having different time profiles, for example, lower-quality solutions needing frequent and costly maintenance against higher-quality alternatives that cost more up front, but promise less frequent and less costly maintenance efforts.

10.8.2 Harmonizing one-time investments and annual benefits

In the OCBA system, all benefits and cost are converted to annuities. For benefits, this is usually not a problem because benefits are already given as annuities: the benefit per year. A one-time investment paid up front is converted into an annuity by dividing it with the annuity factor. With a 3%

discount rate and a project period of 40 years, the annuity factor is 23.81. That is: an up-front investment of €1,000 that resolves into an annuity of €1000/23.81 = €42. This is 68% more than if the discount factor had been zero and the investment distributed over 40 years, giving an annuity of €1,000/40 = €25.

Costs and benefits of a measure or treatment are always calculated relative to a reference situation. This could be a competing alternative or the "zero situation" ("do nothing" or "business as usual"). Because we are focussing narrowly at gains and losses of implementing a measure, it is only necessary to deal with properties that change. Attributes that remain constant won't affect the calculations and can be left out.

Economic analyses come in two flavours: cost-effectiveness analyses (CEA) and cost–benefit analyses (CBAs).

10.8.3 Cost-effectiveness analyses (CEA)

CEA is often employed in situations in which a mandatory environmental limit needs to be reached or where a political/administrative decision has been made to attain a given environmental improvement. The objective is to *minimize* the cost achieving a predefined acoustic goal. Measures having a more efficient design, are more durable, need less maintenance, or employ fewer or cheaper materials, etc., come out on top. (In essence, maximisation alternatives of similar costs are examined, e.g., if there is a fixed budget we may seek the solution that for a given cost provide the largest effects.)

An advantage of CEA is that there is no need to assign a monetary value on the acoustic or environmental target. The "good" provided is measured in "natural units," such as the reduction in dB(A), the reduction in the number of those highly annoyed, etc. CEA, therefore, can be applied in situations where the monetary value of noise impacts has not yet been established. This is currently the case for acoustic improvement of most outdoor and nonresidential situations. Unlike CBA, the result of a CEA does not provide any information concerning the socioeconomic profitability of the alternatives.

The CER of a noise policy can be used to rank alternatives. Given, for example, an objective of reducing the number of highly annoyed persons with a given amount we have:

$$CER = \frac{\text{Cost in } €}{\text{Reduction Highly Annoyed}}. \tag{10.1}$$

You want the CER to be as small as possible.

10.8.4 Cost-benefit analysis (CBA)

We have, given the residential settings, used CBA for the analyses of the noise mitigation measures. CBA takes a more holistic approach than does CEA, by expanding the scope of analysis to all impacts for which those affected in various ways by the measure are willing to pay. For example, noise reduction devices and treatments may provide additional aesthetic, local air pollution, and/or thermal insulation benefits that can be included in the CBA. (High-insulation green roofs can provide microclimatic benefits in the form of less overheating and evaporative cooling and downdrafts[27]. In street canyons, trees and hedges can give protection against wind[28].

The CBA approach is more demanding than is CEA because all relevant effects need to be assigned a monetary value. The expanded scope of CBA may favour the use of more expensive noise reduction devices or treatments in situations where additional benefits are reaped by spending more. Measures that are aesthetically pleasing get a "free" ride by being "subsidised" by aesthetic improvements or other additional benefits.

When undertaking a CBA, we here* use the BCRs:

$$BCR = \frac{\text{Benefits}}{\text{Reduction Highly Annoyed}}. \tag{10.2}$$

If benefits exceed the cost[†] of achieving them, the BCR exceeds one (BCR > 1) and the project is economically "efficient." However, for a noise abatement project to be competitive relative to other projects waiting for public funding, benefits should outweigh the costs by a factor of two or more (BCR > 2). These projects are deemed *robustly* efficient.

When more than one project is being evaluated, they can be ranked in order of profitability using the BCR. The project with the greatest BCR will be considered for implementation first.

10.8.5 Monte Carlo simulations

In Monte Carlo simulations, a large number of CBAs are conducted using input values chosen from each input distribution. The Monte Carlo simulations associate the number of beneficiaries with ±15% uncertainty. We generally use ±20% for the magnitude of noise impacts. We use ±30% valuation of aesthetic/amenity value and for the value of binding CO_2 in

---

* Alternative indicators are the internal revenue rent or net present value (NPV); the latter is robust in terms of varying specifications of benefits and costs (e.g., if an effect of the measure is specified as a cost or a negative benefit/disbenefit, whereas BCR is not, but BCR is better in terms of comparing measures/projects of different sizes.
† Note that an efficient project has a high CBR. An effective measure has a low CER. The two terms are easily confused.

growing trees. Cost estimates are associated with ±20% uncertainty except for wood cuttings, which is part of the pollarding cost where we already have assigned an uncertainty.

10.9 CHARTING THE UNKNOWN: AESTHETIC APPRECIATION

10.9.1 We choose to include aesthetic benefits in the economic considerations

A challenge when undertaking CBA is that there are often some effects for which there are no readily available monetary valuations. Many effects are "public goods" or "nonmarket goods"—effects for which there is no market and, thus, no market prices.

One philosophy is to deal with these factors separately from the economic analyses. This has the advantage of being able to reason from a broader set of premises than just economics. However, ignoring economics carries a real risk of being a disservice to the communities affected by noise. It is human to err in the direction of emphasising immediate costs over vaguely described benefits attained in some distant future.[29] Replacing economic considerations with a discursive exposition of pros and cons is not necessarily a good idea. There is no guarantee that additional nonmonetary considerations will receive the same attention or be given the same emphasis as economic (monetised) considerations that are built into the decision-making framework and tools that are used by the authorities.

Moreover, economic values are based on peoples' willingness to pay for some effect or good, irrespective of transaction or not in markets.[25] It seems obvious, based on the literature and mere common sense, that some people (though not necessarily all people) will pay for a quieter residence or for more greenery near their urban home (if there is not already much greenery in the area). Disregarding such (monetised) benefits from a CBA will make the analysis partial[30] and it might render the CBA less informative or even misleading.

Based on our experience with assessing human reactions and preferences, including those that are intricate to quantify,[7,31–39] we have chosen the more adventurous path of charting the unknown; we have attempted to establish some initial valuations in the form of preliminary unit values for the amenity/aesthetic impact of noise-control measures based on vegetation.[5,10] These unit values provide a starting point for application in CBA and a stepping stone for further analyses. By incorporating these values into the economic calculations, we prepare the ground for a discussion of the relative merit of different effects, and, hopefully, an iterative process where the initial estimates are elaborated and refined over time.

10.9.2 The aesthetic/amenity value of urban greenery

The original amenity/aesthetic valuation estimates stem from studies applying the hedonic pricing method. In hedonic pricing, the sale prices of properties having different characteristics in a market are analysed to estimate the impact of property characteristics on the sales price.[40,41] When data also include information about urban greenery on the property or adjacent to the property, an economic value of the greenery can be derived from the hedonic regression analysis of the properties with different amounts of urban greenery as well as other property characteristics. Our basic approach for deriving initial unit estimates for the ambient/aesthetic benefits of urban greenery is to extract valuation estimates from the literature and relate these estimates to the reported size of the urban greenery. In several studies, no exact measure of the urban greenery area, or size, is reported. Then, we estimate this area from other indicated measures, e.g., number of trees and average canopy size, or the size of roofs or walls that are green.[10] The value estimates for urban greenery from several hedonic pricing studies were plotted onto a spreadsheet, together with the country/city where the data of the study were collected, the year of the data collection, and the currency of the country. We, then, first convert the original urban greenery valuation estimate to Euros, including a CPI (consumer price index) adjustment to Euro 2010 prices (€_{2010}). For the reported property values from the studies, or the urban greenery share of property values, we apply a discount rate of 5% and a 50-year lifetime to calculate annuity values that are comparable to annual rents.[42,43] We obtain the unit value estimate, in €_{2010} prices, by dividing the estimated total urban greenery valuation (per dwelling) by the estimated urban greenery area. This yields an annual value of 1 m² urban greenery for a dwelling (a household).

10.9.3 Valuations studies of vegetated walls/roofs

Peck[44] stated a value increase interval for a property with green roofs or green walls of 6–15%, the midpoint being 10.5%, in Toronto (Canada). For a Toronto house price of approximately CAD 230,000 in 1999, an annualised square metre value of green roofs and walls can be calculated at approximately €_{2010} 20. This can be considered the unit value accruing to one dwelling or household over one year.[10] Des Rosiers et al. estimated that green walls (or hedges) added 3.9% to the property value, in their hedonic price study applying property data from the city of Québec (Canada),[45] where the average price was CAD 112,000 in 1999, yielding a unit value below €_{2010} 4.[10]

Gao and Asami made estate agents assess market values of dwellings in Tokyo and in Kitakyushu (Japan), including the contribution from greenery quality, with scores of –1 for mostly concrete block walls, 0 for average greened walls, and +1 for continuously greened walls.[46] Based on the pricing from the estate agents, they found that an increase in greenery quality level would increase land price by 1.4% in Tokyo (with average property value of about 600,000 JPY in 1999) and by 2.7% in Kitakyushu (with average property value of about 75,000 JPY in 2003). By quantifying the level 0 to 25 m^2 green wall and level +1 to 50 m^2, the resulted unit value estimates are, respectively, €$_{2010}$ 3.5 per square metre green wall in Tokyo and €$_{2010}$ 1 in Kitakyushu. These estimates, of course, are very sensitive to the average wall area assumption for levels 0 and +1.[10]

Ichihara and Cohen found that homes in New York with green roofs had 16.2% higher price than homes without green roofs.[47] Relative to an annual rental price of USD 4,000 in the year 2000, a unit value per square metre green roof of about €$_{2010}$ 17 can be estimated.[10] Tomalty and Komorowski presented value estimates for two types of green roof, a "productive" rooftop garden (including vegetables/fruit) and a "recreational" rooftop garden.[48] They found a 7% and a 20% property price increase in Toronto (with average property value of just below 400,000 CAD in 2010), respectively, for productive rooftop gardens and recreational green roofs. The unit value estimates are about €$_{2010}$ 22 and €$_{2010}$ 63, respectively, per square metre productive rooftop garden and per square metre recreational rooftop garden.

10.9.4 Unit value €2010 2.4 per person per year per square metre wall/roof

The mean of the above referred green wall and green roof value estimates is €$_{2010}$ 18.7 per square metre greenery per household per year (dropping to €$_{2010}$ 12.3 if the highest estimate is removed). Veisten et al.[10] present a weighted mean of the unit value estimates, taking into account the uncertainty in the estimates and yielding higher weight to estimates with lower (reported or calculated) uncertainty. The weighted mean was €$_{2010}$ 1.2; it was driven by the Kitakyushu estimate,[46] which was low and having a tight confidence interval (obtaining almost 90% of the weight). Removing this observation increased the weighted mean per household to €$_{2010}$ 5.8. If this is taken as a "best estimate," then the unit value per person per year, for a square metre green roof/wall, is about 2.4 €$_{2010}$. We stress that this is a preliminary estimate, based on a few studies from different parts of the world.

10.9.5 Valuations studies of urban trees

Morales[49] presented a hedonic valuation of "good" versus "poor" tree cover on properties in Manchester/Hartford, Connecticut. The house price increase was USD 2,686 (1985 USD), or 6%, from "bad" to "good" tree cover, yielding an annuity value of nearly €2010 350. If we quantify "good" to 10 trees and "poor" to zero trees, based on photos in the article, and assume an average canopy size of 25 m^2,[5] again based on visual impressions from photos, the estimated annual square metre value of tree canopy for the household is approximately €2010 1.40, or €2010 35 per tree. Anderson and Cordell[50] found that landscaping with, on average, five (mostly) large trees in the front yard was associated with about 4% increase in property sales prices, or USD 2,970 (1985 USD), in Athens, Georgia. The annuity value for the five trees is calculated to be €2010 295. Assuming a total canopy size of 200 m²,[5] we get an annual square metre value of tree canopy for the household of nearly €2010 1.50, and nearly €2010 75 per tree. Martin et al.[51] presented a study of price effects from trees on properties in Austin, Texas, finding an average house price increase of USD 2,686 (1985 USD). The annuity value for 20 trees is calculated to be €$_{2010}$ 350. Based on statements of "[t]rees were generally small" and "high density of stems" (p. 73), we assume an average canopy size of 25 m²,[5] yielding an annual square metre value of tree canopy for the household of approximately €$_{2010}$ 3.15, or nearly €$_{2010}$ 80 per tree.

Peck[42] stated that "good tree cover" would yield a property increase of 6–15%, thus a midpoint of 10.5% (as for green walls/roofs). For a Toronto house price of approximately CAD 230,000 in 1999, assuming "good tree cover" is five large trees,[5] we get an annual square metre value of tree canopy for the household of approximately €$_{2010}$ 5, or nearly €$_{2010}$ 130 per tree. Des Rosiers et al.[45] estimated that an increase of tree cover between neighbouring properties by 1% would add 0.2% to the property price in Québec City. Assuming that this neighbourhood area is an area with 50 m radius, the 0.2% increase would amount to a 36-m² canopy size.[5] Then, we obtain an annual square metre value of tree canopy for the household of slightly less than €$_{2010}$ 0.3, or €$_{2010}$ 7 per tree. Thériault et al.[52] also applied data from Québec City, estimating the impact from a wooden lot with mature trees to be between −9 and +15% of the property price. Taking the midpoint of 3% and assuming an average tree number of 10, with average canopy size of 75 m²,[5] an annual square metre value of tree canopy for the household of €$_{2010}$ 0.2 is obtained, or €15 per tree.

Stigrall and Elam[53] applied property value data from Lubbock, Texas, where the average property price was USD 162,153 (2006 USD). They found that increasing tree cover from 10 to 25% would increase house values by 3.2% on average, or USD 5,197. Based on the average property/lot size, this increase is equivalent to approximately 90 m² canopy,[5] yielding an annual value of approximately €$_{2010}$ 244, a square metre value of tree canopy of €$_{2010}$ 2.70 (and, assuming that the 90-m² canopy is given by three trees with average canopy size of 30 m², €80 per tree). Dimke[54] valued trees on properties in Cincinnati, Ohio, and found that the trees contributed 10.7% of the property values (USD 188,730 on average) and that 1% increase in tree cover increased prices by USD 784. The average canopy cover was 25.8%. It is stated that the acre value from the hedonic regression is close to the property value; thus, we simply assume that one acre (ca. 4,047 m²) is the average size of the property, although the property price also includes the value of buildings. We assume further that the buildings cover about 30% of the property, yielding an estimate of nearly 15 large trees, each with 50 m^2 canopy area,[5] which fits fairly well to the photo sample of the properties. This yields an annual value for the household of approximately €$_{2010}$ 850 and a square metre value of tree canopy of €$_{2010}$ 1.15 (and about €$_{2010}$ 60 per tree). Wachter and Gillen[55] assessed valuation of urban tree plantings, with data from Philadelphia, Pennsylvania, where the price of average property was USD 82,700 (2004 USD). Urban tree plantings were found to increase property prices by 9% for those having these trees adjacent to their properties (say, within 10 m). Assuming an average number of 20 trees, with a total canopy of 500 m^2, we calculate an annuity value of nearly €$_{2010}$ 370; and assuming an average canopy size of 25 m^2,[5] we get an annual square metre value of tree canopy for the household of about €$_{2010}$ 0.75 (or €$_{2010}$ 18 per tree).

Donovan and Butry[56] valued street trees in Portland, Oregon, and found that an average street tree with 29 m^2 crown area (within a distance of 30.5 m from the property) added USD 19,958/7.6 to sales prices; i.e., 7.6 houses, in average, benefitted from the street tree assuming a threshold distance for obtaining the benefit of 30.5 m (100 feet). This yields annual values of nearly €$_{2010}$ 125, or about €$_{2010}$ 4.25 per m^2. Sander et al.[57] valued urban tree cover in the St. Paul area in Minnesota (Dakota and Ramsey counties). They found that the increase in the home price would be USD 1,371 for 10% increase in tree cover within 100 m. The average property price was USD 287,637 (assumedly USD 2006 values), having an average lot size of 1,375 m^2 (0.34 acres). The average number of street trees within 100 m was 14.55, and the average tree cover on the properties was 15.44%. Assuming an average canopy size of 25 m^2,[5] we obtain annual values for the household of about €$_{2010}$ 500 for the street trees within 100 m, with a canopy square metre value of €$_{2010}$ 1.40 (or €35 per tree).

10.9.6 Unit value €2010 0.50 per person per year per square metre canopy

The mean of the urban tree unit value estimates is about €$_{2010}$ 2.3 per square metre canopy, per household per year. The interval from the 5% percentile to the 95% percentile goes from about €$_{2010}$ 0.25 to about €$_{2010}$ 4.6. The average value estimate per tree was approximately €$_{2010}$ 75 (the 5% percentile to the 95% percentile goes from about €$_{2010}$ 10 to about €$_{2010}$ 170). The weighted mean of the unit value estimates, taking into account the reported (or calculated) uncertainty in the original studies, is €$_{2010}$ 1.15 per square metre tree canopy per household per year, or about €0.60 per person per year. This result is, to a large extent, driven by the Québec estimate from Des Rosiers et al.;[45] if this is removed, the weighted mean increases to about €$_{2010}$ 2 per household (ca. €$_{2010}$ 1 per person per year). Thus, applying €$_{2010}$ 0.50 per person per year per square metre canopy, for urban trees within a maximum of 100 m from the residence, seems a fairly conservative estimate.

10.10 CONCLUDING REMARKS

This chapter provides economic calculations of different classes of environmental noise mitigation measures. The utility of assessing in a systematic and uniform manner benefits and costs of environmental measures to reduce noise has been demonstrated. Leaner products, improved use of materials, recycling, appropriate maintenance, longer life cycles, and profiting from "side effects" mean that more people may benefit from scarce resources available for improving noisy urban environments.

For the CBA results to be valid, it is necessary that all substantial effects, also nonacoustic, are taken into account. Not only improvements in sound quality and reductions of noise need to be assessed, but also aesthetic, amenity, and other benefits. The unit values presented here are based on relatively few studies, but provide a starting point for testing and further analyses.

Commercial products are put into market for a profit. Therefore, there are clear incentives to improve the overall qualities of the product relative to their costs. Public goods are often brought into place by different authorities or parties planning and organising their activities in a more fragmented manner, without the same considerations to maximising overall benefits and keeping overall costs down. There, thus, is a need for authorities to not only view themselves as users or purchasing agents of environmental measures to reduce noise, but also as contributors to the tool chest available for European agglomerations, roads, and rail authorities in reducing noise. This they can do if, in every project of any size, there is set aside a sum for control measurements and where the overall performance of the measures is evaluated.

Cities are increasingly aware that they need to pay attention to the qualities of the urban areas to be competitive, attracting businesses and a highly qualified workforce, and neighbourhoods where it is both good to live and bring up children. Here, the overall quality plays a role. The quality of the soundscape also should be considered when developing the urban landscape and the blue and green areas of the city.

REFERENCES

1. Klæboe, R. 2012. CBA framework with updated cost and benefit estimates (FP7 HOSANNA Deliverable 7.2). Online at: http://www.greener-cities.eu/
2. Bickel, P., R. Friedrich, A. Burgess, P. Fagiani, A. Hunt, G. de Jong, J. Laird, C. Lieb, G. Lindberg, and P. Mackie. 2006. Proposal for harmonised guidelines. Deliverable 5: Developing Harmonised European Approaches for Transport Costing and Project Assessment (HEATCO). Stuttgart: Institut für Energiewirtschaft und Rationelle Energieanwendung (IER).
3. Andersson, H., and M. Ogren. 2013. Charging the polluters: A pricing model for road and railway noise. *Journal of Transport Economics and Policy (JTEP)* 47 (3): 313–333.

4. Nijland, H. 2008. Theory and practice of the assessment and valuation of noise from roads and railroads in Europe (Ph.D. thesis). Technical University Delft.

5. Veisten, K., R. Klæboe, and M. Mosslemi. 2013. Valuation of urban trees from hedonic price studies of property sales data: A literature survey with meta-analysis. Unpublished paper.

6. Dol, K., and M. Haffner. 2010. Housing statistics in the European Union. The Hague, The Netherlands: Ministry of the Interior and Kingdom Relations, p. 150.

7. Klæboe, R. 2011. Noise and health: Annoyance and interference. In *Encyclopedia of environmental health*, Vol. 3, ed. J. O. Nriagu (pp. 152–163). Burlington, MA: Elsevier.

8. De Kluizenaar, Y., E. M. Salomons, S. A. Janssen, and F. Van Den Berg. 2012. Traffic noise and annoyance: The effect of quiet façades and quiet areas. Paper presented at the Proceedings of the European Conference on Noise Control, Prague.

9. Salomons, E., H. Zhou, H. Vos, Y. De Kluizenaar, S. Janssen, and F. Van Lenthe. 2010. Model study of urban traffic noise and quiet facades, in relation with annoyance and sleep disturbance. In *Noise in the built environment*, Vol. 32 (pp. 359–366). Ghent: Institute of Acoustics.

10. Veisten, K., Y. Smyrnova, R. Klæboe, M. Hornikx, M. Mosslemi, and J. Kang. 2012. Valuation of green walls and green roofs as soundscape measures: Including monetised amenity values together with noise-attenuation values in a cost-benefit analysis of a green wall affecting courtyards. *International Journal of Environmental Research and Public Health* 9 (11): 3770–3778.

11. Van Den Berg, F., C. Schoonebeek, and M. Hillebregt. 2012. On the definitions quiet façade and quiet Urban areas. Paper presented at the Proceedings of the European Conference on Noise Control, Prague.

12. Amundsen, A. H., R. Klæboe, and G. M. Aasvang. 2011. The Norwegian façade insulation study: Efficacy of facade insulation in reducing noise annoyance due to road traffic. *Journal of the Acoustical Society of America* 129 (3): 9.

13. Amundsen, A. H., G. M. Aasvang, and R. Klæboe. 2013. Long-term effects of noise reduction measures on noise annoyance and sleep disturbance: The Norwegian facade insulation study. *JASA* 133 (6).

14. Van Renterghem, T., and D. Botteldooren. 2012. Focused study on the quiet side effect in dwellings highly exposed to road traffic noise. *International Journal of Environmental Research and Public Health* 9 (12): 4292–4310.

15. CityHush. 2013. Acoustically green road vehicles and city areas. Stockholm, project home page.

16. Bashir, I., S. Taherzadeh, and K. Attenborough. 2013. Diffraction assisted rough ground effect: Models and data. *Journal of the Acoustical Society of America* 133 (3): 1281–1292.

17. Attenborough, K., I. Bashir, and S. Taherzadeh. 2011. Outdoor ground impedance models. *Journal of the Acoustical Society of America* 129 (5): 2806–2819.

18. Bashir, I., S. Taherzadeh, and K. Attenborough. 2010. Diffraction assisted rough ground effect: models and data. Paper presented at the 160th Meeting of the Acoustical Society of America, November 15–19, Cancun, Mexico, Department of Design, Development, Environment and Materials.

19. Van Maercke, D., T. Leissing, and J. Maillard. Technical report for task 6.2 (FP7 HOSANNA Deliverable 6.2). Online at: http://www.greener-cities.eu/

20. Canevaflor. 2013. Vegetated walls. Online at: http://www.canevaflor.com/page/340

21. Combes, P.-P., G. Duranton, and L. Gobillon. 2013. The costs of agglomeration: Land prices in French cities. In Discussion papers. London: Centre for Economic Policy Research.

22. Van Renterghem, T., D. Botteldooren, and K. Verheyen. 2012. Road traffic noise shielding by vegetation belts of limited depth. *Journal of Sound and Vibration* 331 (10): 2404–2425.

23. Boardman, A., D. Greenberg, A. Vining, and D. Weimer. 2001. *Cost-benefit analysis, concepts and practice.* Upper Saddle River, N.J.: Prentice Hall.

24. Brent, R. J. 1996. Applied cost-benefit analysis. Cheltenham, U.K.: Edward Elgar.

25. Mishan, E. J. 1988. *Cost-benefit analysis: An informal introduction*, 4th ed. London: Uniwin Hyman.

26. Dasgupta, A. K., and D. W. Pearce. 1975. *Cost-benefit analysis: Theory and practice.* London: Macmillan.

27. Susca, T., Gaffin, S.R., Dell'Osso, G.R. 2011. Positive effects of vegetation: Urban heat island and green roofs. *Environmental Pollution* 159 (8–9): 2119–2126.

28. Willemsen, E., Wisse, J.A. 2007. Design for wind comfort in The Netherlands: Procedures, criteria and open research issues. *Journal of Wind Engineering and Industrial Aerodynamics* 95 (9–11): 1541–1550.

29. Kahneman, D., and A. Tversky. 1979. Prospect theory: Analysis of decision under risk. *Econometrica* 47 (2): 263–291.

30. Moore, T., and R. Pozdena. 2004. Framework for an economic evaluation of transportation investments. In *Economic impacts of intelligent transportation systems: Innovations and case studies.* Research in transportation economics, Vol. 8. Amsterdam: Elsevier.

31. Amundsen, A. H., G. M. Aasvang, and R. Klæboe. 2010. Long-term effects of façade sound insulation on noise annoyance and sleep disturbances. Paper presented at the 39th International Congress on Noise Control Engineering—Internoise 2010, Lisbon. Portuguese and Spanish Acoustical Societies, Vol. CD, p. 672, eds. J. Patrício and A. Pérez-López.

32. Fyhri, A., and R. Klæboe. 2009. Road traffic noise, sensitivity, annoyance and self-reported health-A structural equation model exercise. *Environment International* 35 (1): 91–97.

33. Klæboe, R., A. H. Amundsen, and A. Fyhri. 2008. Annoyance from vehicular air pollution: A comparison of European exposure-response relationships. *Atmospheric Environment* 42 (33): 7689–7694.

34. Amundsen, A. H., R. Klæboe, and A. Fyhri. 2008. Annoyance from vehicular air pollution: Exposure-response relationships for Norway. *Atmospheric Environment* 42 (33): 7679–7688.

35. Klæboe, R., E. Engelien, and M. Steinnes. 2006. Context sensitive noise impact mapping. Applied Acoustics 67: 620–642.

36. Fyhri, A., and R. Klæboe. 2006. Direct, indirect influences of income on road traffic noise annoyance. *Journal of Environmental Psychology* 26: 27–37.

37. Klæboe, R., M. Kolbenstvedt, A. Fyhri, and S. Solberg. 2005. The impact of an adverse neighbourhood soundscape on road traffic noise annoyance. *Acta Acustica United with Acustica* 91: 1039–1050.
38. Klæboe, R., A. H. Amundsen, A. Fyhri, and S. Solberg. 2004. Road traffic noise: The relationship between noise exposure and noise annoyance in Norway. *Applied Acoustics* 65: 893–912.
39. Klæboe, R., I. H. Turunen-Rise, L. Harvik, and C. Madshus. 2003. Vibration in dwellings from road and rail traffic. Part II: exposure-effect relationships based on ordinal logit and logistic regression models. *Applied Acoustics* 64: 89–109.
40. Choumert, J., and J. Salanié. 2008. Provision of urban green spaces: Some insights from economics. *Landscape Research* 33 (3): 331–345.
41. Freeman, A. M., and R. F. T. Future. 1979. *The benefits of environmental improvement: Theory and practice.* Baltimore, MD: Johns Hopkins University Press.
42. Sælensminde, K., and K. Veisten. 2005. Report No. SILVIA document SILVIA-TOI-003-03-WP3-310305, p. 1.
43. Nelson, J. 2008. Hedonic property value studies of transportation noise: Aircraft and road traffic. In *Hedonic methods in housing markets*, eds. A. Baranzini, J. Ramirez, C. Schaerer, and P. Thalmann (pp. 57–82). New York: Springer.
44. Peck, S. W., Callaghan, C., M. E. Kuhn, and B. Bass. 1999. Greenbacks from green roots: Forging a new industry in canada. Report. Peck & Associates.
45. Des Rosiers, F., M. Thériault, Y. Kestens, and P. Villeneuve. 2002. Landscaping and house values: An empirical investigation. *Journal of Real Estate Research (JRER)* 23: 139–160.
46. Gao, X., and Y. Asami. 2007. Effect of urban landscapes on land prices in two Japanese cities. *Landscape and Urban Planning* 81: 155–166.
47. Ichihara, K., and J.P. Cohen. 2011. New York City property values: What is the impact of green roofs on rental pricing? *Letters in Spatial and Resource Sciences* 4: 21–30.
48. Tomalty, R., and B. Komorowski. 2010. Curb appeal. *Living Architecture Monitor* 12: 26–27.
49. Morales, D. J. 1975. *The contribution of trees to residential property value, Manchester, Connecticut.* Storrs, CT: University of Connecticut.
50. Anderson, L. M., and H. K. Cordell. 1988. Influence of trees on residential property values in Athens, Georgia (U.S.A.): A survey based on actual sales prices. *Landscape Urban Plan* 15 (1–2): 153–164.
51. Martin, C. W., R. C. Maggio, and D. N. Appel. 1989. The contributory value of trees to residential property in the Austin, Texas metropolitan area. *Journal of Arboriculture* 15 (3).
52. Thériault, M. 2002. The impact of mature trees on house values and on residential location choices in Quebec city. Paper presented at the Conference of the International Environmental Modelling and Software Society, June 24–27, Lugano, Switzerland (unpublished).
53. Stigrall, A., and E. Elam. 2009. Impact of improved landscape quality and tree cover on the price of single-family homes. *Journal of Environmental Horticulture* 27 (1): 24–30.
54. Dimke, K. C. 2008. Valuation of tree canopy on property values of six communities in Cincinnati (Ph. D. thesis.) Columbus, OH: Ohio State University.

55. Wachter, S., and K. Gillen. 2006. *Public investment strategies: How they matter for neighborhoods in Philadelphia: Identification and analysis.* Philadephia, PA: Pennsylvania Horticultural Society.

56. Donovan, G. H., and D. T. Butry. 2010. Trees in the city: Valuing street trees in Portland, Oregon. *Landscape Urban Plan* 94 (2010): 77–83.

57. Sander, H., S. Polasky, and R.G. Haight. 2010. The value of urban tree cover: A hedonic property price model in Ramsey and Dakota Counties, Minnesota, USA. *Ecological Economics* 69 (8): 1646–1656.

Index

A

Printed and bound by CPI Group (UK) Ltd, Croydon, CR0 4YY

01/11/2024

01782624-0004